The Truth Society

EXPERTISE

CULTURES AND
TECHNOLOGIES
OF KNOWLEDGE

EDITED BY DOMINIC BOYER

A list of titles in this series is available at cornellpress.cornell.edu

The Truth Society

Science, Disinformation, and Politics in Berlusconi's Italy

Noelle Molé Liston

Cornell University Press
Ithaca and London

First published 2020 by Cornell University Press

Library of Congress Cataloging-in-Publication Data

Names: Liston, Noelle Molé, author.
Title: The truth society : science, disinformation, and politics in Berlusconi's Italy / Noelle Molé Liston.
Description: Ithaca [New York] : Cornell University Press, 2020. | Includes bibliographical references and index.
Identifiers: LCCN 2020003687 (print) | LCCN 2020003688 (ebook) | ISBN 9781501750786 (hardcover) | ISBN 9781501750793 (paperback) | ISBN 9781501750809 (ebook) | ISBN 9781501750816 (pdf)
Subjects: LCSH: Political culture—Italy. | Truthfulness and falsehood—Political aspects—Italy. | Communication in politics—Italy. | Science—Italy—Public opinion. | Fake news—Political aspects—Italy. | Mass media—Political aspects—Italy. | Knowledge, Sociology of. | Italy—Politics and government—1994-2018. | Italy—Politics and government—2018-
Classification: LCC JA75.7 .L636 2020 (print) | LCC JA75.7 (ebook) | DDC 306.20945—dc23
LC record available at https://lccn.loc.gov/2020003687
LC ebook record available at https://lccn.loc.gov/2020003688

Dedicated to Lane and Everheart

It used to be, everyone was entitled to their own opinion, but not their own facts. But that's not the case anymore.

—Stephen Colbert

Give a man a mask and he'll tell you the truth.

—Oscar Wilde

Only in Italy can truth be sold in a political market and ushered into public discourse like a banal fiction.

—Massimo Giannini, *La Repubblica*

Italian information is decomposing. It's like a journalism of new monsters that are eating our brains. . . . News is the triumph of the undead.

—Beppe Grillo, *Il Blog di Beppe Grillo*

Contents

List of Illustrations ix
Preface xi
Acknowledgments xiii
List of Abbreviations xvii

Introduction 1
1. Manifest Disguise and Mediatized Politics 31
2. The Soldiers of Rationality 52
3. The Rise of Algorithm Populism 73
4. The Trial against Disinformation 96
5. Scientific Anesthetization in the Anthropocene 120

Conclusion: Mirrored Window World 142

Notes 165

Bibliography 181

Index 207

Illustrations

Figures

0.1. *Jettatore* in piazza, June 2011. 5
0.2. Venice graffiti, June 2017. 19
1.1. Italian prime minister Silvio Berlusconi in Milan in April 2010. 33
1.2. News parody program *Striscia*'s Golden Tapir Award (Tapiro d'Oro). 44
1.3. News parody program *Striscia*'s puppet Gabibbo in Turin in 2008 at an event for Italian Union of Parents against Children's Tumors (Unione dei genitori contro i tumori dei bambini, or UGI). 47
2.1. CICAP's "A Day against Superstition," June 2011. 53
2.2. A participant in "A Day against Superstition" breaks a mirror, June 2011. 64
2.3. CICAP's main table featuring the tally of dice rolls at CICAP's "A Day against Superstition," June 2011. 66

2.4. The guesses at dice rolls before and after engaging in a superstitious act at CICAP's "A Day against Superstition," June 2011. 67
3.1. Beppe Grillo in Frascati, Italy, at Five Star Movement rally, 2009. 76
4.1. Antonio gazes at a map of L'Aquila, June 2013. 98
4.2. A street in the abandoned city center of L'Aquila, June 2013. 115
C.1. Image of Luigi di Maio, 2019. 155

Preface

What is your *one question*—the question all your scholarly work interrogates? I was asked this on a job interview on the car ride between campus interviews and dinner. I answered that it was the connection between knowledge and embodied experience. How does what we know shape what we feel? How does what is within the available knowable reality, which is always so highly contingent on cultural and historical time and place, shape how we make sense of the world through to our deepest somatic registers? My work seems to continually hover around the kinds of things we might experience as most intimate and untethered, then tries to move outward, curious about what knowledge—emergent, institutionalized, structural—frames and molds and sustains it.

My first book, *Labor Disorders in Neoliberal Italy* (2011), examined "mobbing," a term in Europe and Italy to name workplace harassment, exclusionary and isolating behaviors that typically forced workers into quitting. At the time of my work, mobbing appeared to be proliferating—as evidenced by new counseling centers dedicated to help "victims of mobbing"; a work-related illness named as a result of mobbing; new occupational laws, regulations, and best practices; new human resources training; and news stories and media dedicated to mobbing. Why was mobbing suddenly capturing the attention of Italians? I found it especially fascinating that this new mounting knowledge about mobbing was disseminated so rapidly and persuasively that a person would understand their own stomachache as "mobbing sickness" or look at a colleague and think, "mobber." Understanding why such naming was possible, I argued, emerged from a long history of safeguards and cultural expectations about work and longevity, rapid neoliberalizing in the 1990s and 2000s, and an increasing sense of precarity. To identity as a victim of mobbing was wrapped up in a whole constellation of knowledges and structures.

This project began when I heard about national events such as Unified Italy for the Correct Scientific Information (Italia Unita per la corretta informazione scientifica) with the hashtag "Italy4science." One panel of scientific speakers in Milan was supported by the Democratic Party (Partito Democratico) and European Federalist Radical (Radicale-Federalista Europeo) Party and Left Ecology and Freedom Party (Sinistra Ecologia e Liberta'). Why did science need defending? And why was such an event for "correct information" sponsored by left-wing political parties? In 2013, I heard about the work of the Committee for the Investigation of Pseudoscientific Claims (Comitato Italiano per il Controllo delle Affermazioni sulle Pseudoscienze, CICAP) and their Day against Superstition (Una Giornata Anti-superstizione). Why would people want to protest superstition? At the event, I watched as members of CICAP earnestly tried to persuade Italians that black cats, mirror breaking, and salt spilling had no power to bring them misfortune or illness. They encouraged participants to throw the salt and aimed to prove, with the power of a statistical dice game, that superstition was a flawed logic. In this book, I build toward my fieldwork with CICAP in the last chapter (chapter 5) even though it was this very mystery that launched *The Truth Society*. But that image—of CICAP activists trying so earnestly to encourage salt spilling and persuade onlookers of its insignificance and convert them to proper rational understanding of salt—struck me deeply. Why did salt tip from a silly measure of predicting misfortune to a dangerous sign that fellow Italians were believing in false truths? Once again, I began an intellectual journey to understand how particular kinds of knowledge about the world were part of reimagining this familiar ritual, and how seemingly unrelated political and technological material change reshaped its stakes.

Acknowledgments

I began fieldwork for this project during my years at the Princeton Writing Program and am grateful for funds from the Princeton University Committee on Research in the Humanities and Social Sciences. At New York University, I am grateful for my research budget through the Expository Writing Program.

I am indebted to Angelique Haugerud, Dominic Boyer, and my anonymous reviewers for their insightful feedback on chapter 1; *American Ethnologist*'s Linda Forman for smart editing work and enthusiasm; Sadi and Nader Marhaba for research support; and photographers Roberto Di Cristina, Bruno Cordioli, and Antonio Cardinale for permission to use their work. I am grateful to *American Ethnologist* for permission to republish significant portions of the article "Trusted Puppets, Tarnished Politicians: Humor and Cynicism in Berlusconi's Italy" here. I am also grateful for permission from the Council on European Studies, publishers of *European Politics and Society,* for allowing reprinting of my work, "Enchanting the Disenchanted: Grillo's Supernatural Humor as Populist Politics," as well as to Neringa Klumbyte for her smart suggestions and conversation on the piece. I have presented this work over years at numerous conferences, including American Anthropological Association meetings, American Ethnological Society meetings, the Society for Cultural Anthropology, PSI: Performance Studies International, the 2011 Barcelona meetings of the Society for the Anthropology of Europe and Council for European Studies, and a 2018 conference on politics and humor, the "Comedy World Summit," with special thanks to Morten Nielsen. I have received vital feedback from collaborators and colleagues that has changed and spurred my thinking. All translations from Italian to English are mine.

My family has supported me and made my scholarly life richer and easier: my mother, Maureen Molé, and Marissa and Paul Bostick; Dana and Nick Flynn; Maj. Lane L. Liston Jr. and Caroline Liston; Esther and Stephen Winikoff; my late grandparents, Carmel and Vincent Altomare. I am very grateful to many conversations with my fellow Italianist, dear friend, and brilliant scholar Andrea Muehlebach. I am also grateful for formative and inspiring conversations with fellow scholars of Italy over these years: Betsy Krause, Tracey Heathington, Emanuela Guano, Lilith Mahmud, Jason Pine, Jillian Cavanaugh, Pamela Ballinger, and Stavroula Pipyrou.

I am greatly indebted to artist Simona Conti for use of her painting and the fabulous image of it by friend and photographer Nikki Alcazer. I have been enriched over the years through lively dialogue with Mark Robinson, Leo Coleman, Mona Bhan, and Jennifer Karlin; my New York University crew that keep my mind engaged, Elena Glasberg, Jennifer Cayer, Katherine Carlson, Jacqueline Reitzes, Jenni Quilter, Leah Souffrant, Ben Stewart, and Doug Dibbern. Special thanks to EWP colleague and poet Mara Jebsen for wordsmithing my book title.

I am also greatly indebted to the lucid insights from my manuscript reviewers. I am immensely grateful for the hard work of the entire Cornell University team, Jim Lance, Clare Kirkpatrick Jones, and Brock Schnoke; and for the excellent scholarship and guidance of series editor, Dominic Boyer. I also thank Amron Gravett of Wild Clover Book Services for indexing.

I am deeply thankful for all my interlocuters in Italy who so generously shared their time and their stories with me.

My circle of friends in Italy has supported me in so many ways over the years, as kind and generous as they are fierce social analysts: Francesca Previati and Riccardo Panzarini; Margherita Masignani and Alessandro Minin;

Grazia Morra; Sadi, Nader, and Shadia Marhaba; William Murphy and Giovanna Tomasi; Diego Vertieri, George and Sandy Basmaji; Giada Marini, and Ruggero and Patrizia Falconi. I am especially grateful to the late Barbara Falconi, to whom I owe an enormous debt, as she became my very first Italian friend when she was my "host mother" during my undergraduate semester abroad in Florence. She shared her world with me, a magnificent ocean of artists and musicians, sailors and mariners, misfits and outcasts. My journey toward cultural anthropology and to Italian studies was ignited in Barbara's gift of inclusion, and her own awe-inspiring way of living against the grain.

My scholarly work is nourished by the incredible support of my partner, Lane Franklin Liston. And to our son, Lane Everheart Liston, thank you for coming along on our Italian adventures and for all the joy and love you've brought into my world.

Abbreviations

CICAP	Committee for the Investigation of Claims of the Paranormal (Comitato Italiano per il controllo delle affermazioni sul paranormale) and Committee for the Investigation of Pseudoscientific Claims (Comitato Italiano per il Controllo delle Affermazioni sulle Pseudoscienze)
DC	Christian Democrats (Democrazia Cristiana)
DPC	Civil Protection Department (Dipartimento della Protezione Civile)
HAARP	High Frequency Active Auroral Research Program
INGV	Italy's National Institute of Geophysics and Volcanology (Istituto Nazionale di Geofisica e Vulcanologia)
M5S	Five Star Movement (Movimento Cinque Stelle)
PCI	Communist Party (PCI)
PSI	Italian Socialist Party (Partito Socialista Italiana)
PD	Democratic Party (Partito Democratico)
PDL	The People of Freedom Party (Il populo della Libertà)

The Truth Society

Introduction

> We need to unmask what could be called the "snake-tactics" used by those who disguise themselves in order to strike at any time and place. This was the strategy employed by the "crafty serpent" in the Book of Genesis, who, at the dawn of humanity, created the first fake news (cf. Gen. 3:1–15).
>
> —Pope Francis, *Message from His Holiness on World Communications Day*

> Knowledge is not a direct grasp of the plan and the visible . . . but an extraordinarily daring, complex, and intricate confidence in chains of nested transformations of documents that, through many different types of proof, lead away toward new types of vision.
>
> —Bruno Latour, *On the Modern Cult of Factish Gods*

Standing alone under the white tent that shielded him from the summer sun, an elderly man dressed in black from hat to boots looked somberly off into the distance. I recognized him from Neapolitan folklore; he was dressed as a *jettatore*, that is, a "projector" or distributor of bad luck, carrier of the evil eye and bad luck. The wind pushed the edges of the tent around him back and forth; the blue letters of the acronym CICAP were still legible. There were a few tables arrayed with objects not immediately related to one another: boxes of salt, sets of dice, a large easel with an oversized notepad on which were written names and numbers, and an open ladder. A young man at the table invited me to play a dice game, which would, he promised, convince me that any superstitious beliefs I held about these practices were, in fact, not based in reality, only fictions I had come to see as true. A few feet from him, a few dozen had gathered around watching curiously as people smashed mirrors with tiny hammers. It was the inaugural sound of the "Day against Superstition," a demonstration held by CICAP, the Italian Committee for Investigation of Pseudoscientific Claims, in Vicenza, Italy. About the occasion, CICAP founder Massimo Polidoro said, "Since 1989, CICAP has been committed to fighting irrationality, superstition and prejudice with the weapons of science and reason" ("Venerdi 17"). The event was planned on a day known for bringing misfortune and bad luck: Friday the 17th of June, 2011.

Indeed, that day, June 17, 2011, was not a particularly lucky day for Italy's then prime minister, Silvio Berlusconi.[1] Frustrated with a tumultuous corruption scandal over the past several months, he said, "I should liquidate everything and leave Italy" (Bei). When he entered Rome's Palazzo Chigi, the prime minister's official residence, and he was asked about new approval ratings that said he was soon to be ousted, he replied plainly: "They're all lies."

Meanwhile, the university chancellor (*rettore*) of the University of L'Aquila celebrated a new scholarly volume, *The L'Aquila Earthquake: Analysis and Reflection on the Emergency*, on the 2009 earthquake and its repercussions.[2] The university chancellor commented, "After the earthquake I thought: 'Why not put together a team of our professionals that can make observations in an objective manner, that can think through the facts with a technical-scientific approach?" ("Presentatato Il Libro" 2011). Just a few weeks earlier, on May 25, Public Prosecutor Giuseppe Romano Gargarella indicted the Great Risk Commission, a group of scientists responsible for evaluating seismic risks, in an accusation of thirty-two homicide deaths on the day of the earthquake, April 6, 2009. The scientists were accused of issuing "generic and ineffective information in relation to the duty to predict and prevent earthquake, and issuing incomplete, imprecise and contradictory information on the nature, cause, and dangerousness and future developments of seismic activity" ("Terremoto a L'Aquila" 2011). At its meeting on March 31, a week before the earthquake, the Risk Commission had issued "reassuring declarations that induced many Aquilans to stay in their homes" ("Terremoto a L'Aquila" 2011). At a press conference, scientists had told the anxious citizens of the region that earthquake prediction was scientifically impossible but also that the recent swarms, small terrestrial tremors, did not indicate a coming quake. One scientist even recommended citizens enjoy some wine at home. Days later, a 6.3 magnitude earthquake devastated the city of L'Aquila, killing over three hundred people and leaving over sixty thousand people homeless. To the shock of the world's scientific community, the scientists were charged with manslaughter: the courts criminalized their "false reassurances," issued at the news conference, as the cause of certain victims' deaths.

In September of 2011, the case against the Risk Commission would have its first hearing; it was a trial that some called "the Trial against Science" (Ciccozzi 2016).

Beyond the shared day, unlucky Friday the 17th of 2011, there is something curiously resonant across these seemingly disparate events. At a glance, we have a piazza in Vicenza where demonstrators used science "as a weapon" against fellow citizens to undo their supposedly irrational superstitions. In Rome, a prime minister, in his last few months in office after nearly twenty years at Italy's helm, casually dismissed a poll as "lies." And just weeks after scientists in L'Aquila were indicted for homicide for giving citizens reassuring information that would lead to their deaths, a university provost lauded a "technical-scientific" study of historical fact.

There is a quivering thread here that unravels our assumptions about the existence of fact, about how we determine what is true about the world. None of the actors involved in these events took objective facts for granted. Put differently, the notion of definitive truth does not appear to be a shared epistemological and cultural premise. In a country where pagan and Catholic beliefs have, for centuries, allowed for the coexistence of superstition and science, why protest superstition today? Why did facts need defending, or seem, in some way, endangered? Why would scientists become legally culpable for reassuring the public? Why would the public have such profound faith in these scientists' reassurances as to remain in their homes? *The Truth Society* seeks to understand why and how the Berlusconian era of political spectacle, which regularly blurred fact and fiction, shapes how people understand truth, particularly mass-mediated news and information, and in turn, scientific knowledge. Berlusconi's casual dismissal of fact offers us a valuable clue that is the context in which to understand scientific activists and the Aquilan trial. Showing how epistemology is shaped by political life and how political life is shaped by material forms of knowledge, I examine how late twentieth-century Italian politics dramatically fashioned this peculiar coupling in public life between fact and fiction and the seemingly oppositional investment in disinformation and science.

Mapping the epistemology of science as an outgrowth of political life, I scrutinize Italy's late twentieth-century political culture, particularly former prime minister and media mogul Silvio Berlusconi. Known for both his charisma and corruption, Berlusconi rebooted Italy's long tradition of theatrical politics for the twenty-first century with his media savvy, exacerbating a political culture in which facts' packaging trumped their accuracy. The media artifice of the Berlusconi decades has intensified Italian cynicism, producing a public weary of political deception yet nevertheless persuaded by the fabulous fabrication of truth. I track how these shifts are manifested in belief as well as new practices of public activism, media engagement, and the rising populist Five Star Movement. Thus, as a whole, we will find that adherence to and rejection of what is termed "science" represents a culturally particular epistemological and material practice, one shaped by political culture and media as much as by economic uncertainty and volatility.

In perhaps the most basic form, the culture of truthlessness shapes how citizens understand the truth of science. It may have played out differently. In fact, we might have predicted that the rise of political fictions, together with economic uncertainty and intensifying wealth disparity, would give rise to more enchantment: a rise in superstition, nonscientific thinking, magic, and the occult (Comaroff and Comaroff 2000). And while we do see an uptick in enchantment tropes and antiscience conspiracy theories, we also see the strengthening of scientific thinking, a fascination with and engagement with science that amounts to a kind of intensified scientism, a love affair with rationality, even something verging on a kind of credulousness in science (Ciccozzi 2017; Tipaldo 2019). Especially clear in the case of L'Aquila, we find a sort of scientific enchantment or "anesthesia," as one local put it, in which scientific truths might become a new form of sacred and fetishized knowledge, perhaps an overcompensation for its perceived disappearance. The culture of truthlessness may compel people to disproportionately invest in the truth of science and fantasize about a single and rational truth of nature. When I first began doing fieldwork with CICAP in 2011, all I knew was that CICAP's mission, fancied in militarized terms, to disseminate science to the public seemed peculiar. It was only after I began studying the trial in L'Aquila that I realized that CICAP helped me unravel the stakes of scientific truth and managing natural disasters.

Thus, the curious stories of CICAP, L'Aquila, and the Five Star Movement's digital populism hold generative insights for understanding knowledge and belief in the early twenty-first century. They represent a visible ripple of a much deeper crisis shaking the very foundations of knowledge, a crisis that has mangled the process of how people distinguish true from false. The project thus embraces what Andreas Glaeser (2010) calls political epistemology or "the historically specific politics-oriented knowledge-making

Figure 0.1. ***Jettatore*** **in piazza, June 2011. Photo by author.**

practices of people and their consequences" (xxvii). My fundamental premise is that these seemingly disparate stories—the L'Aquila court's legal indictment of scientists' disinformation, its pro-science movement of skeptics, and its cyberutopian political movement—emerge as reactions and responses to a Berlusconian regime of spectacular media and fantastic politics.

Broadly conceived, the culture of semitruths and nontruths changes how people come to understand how truths are made and fabricated, and therefore, perceptions of science as body of knowledge. I consider the audacious act of holding scientists—and media—accountable for poor information in times of profound disinformation as a symptomatic rupture. We must, then, slow down and scrutinize the process of how a political culture of artifice makes some actors doubt science and all forms of authoritative knowledge, while others disproportionately invest in them. In the decades following Italy's first election of a global leader in televised politics, we find the emergence of a disinformation society: a social order characterized by ambivalence regarding how truths are verified in which, of the first magnitude, spectacle eclipses veracity, and, in particular contexts, cautious skepticism and deep suspicion regarding facts emerge. This is a story of how conspiracy and reenchantment, as well as reactionary rationalism, occupy the splits cleaved by Italy's ruptured fact authentication process. Millennial Italy provides an ideal terrain to map these epistemological tremors: its population has long assumed that power—whether Catholic or secular—implies the manipulation of truth, and its leaders have long made power a performance. Drawing together Italy's core institutions—government, courts, and the media—this study investigates how Italy's enchanted political culture shapes how actors invest in science yet also grow increasingly skeptical, even paranoid, about which truths are genuine and who might reliably corroborate them.

From Truth to Post-truth: A Brief History of the Erosion of Facts

In order that we may understand the shift toward radical skepticism about truth, an era in which facts are questioned and objective reality is not shared, I will outline two central contextual trajectories: (1) the philosophy and social theory of knowledge and (2) the historical, within which one story highlights socioeconomic conditions while the other highlights mediatized information. For the former, I will briefly trace how truth became understood as produced and disseminated, which overlaps with but is also distinct from a related field in science studies about how science approximates, represents, and produces empirical reality (Sismondo 1996).

In a *Newsweek* article, lecturer in journalism Andrew Calcutt (2016) blames the age of "post-truth" on Jean-Francois Lyotard's (1984) theory of postmodernism which was, in turn, "shamefully" taken up by "left-leaning, self-confessed liberals [who] sought freedom from state-sponsored truth." Calcutt's declaration is certainly a wild oversimplification; however, this particular intellectual genealogy of ideas is not altogether off-base. Lyotard's (1984) "simplified" understanding of postmodernism was "an incredulity towards metanarratives," that is, unifying and singular theories about the world, including science (xxiv). He did not suggest that people no longer believed in science or "salvation" from economic inequality but rather that there was a spike in the underlying skepticism that sustained belief in these metanarratives, particularly scientific theory and the notion of a single objective truth (xxiv). Belief in science, therefore, was bolstered and "legitimated" by other kinds of knowledge or narratives (7), as well as the state's credibility (28), such that a "crisis of scientific knowledge . . . represents an internal erosion of the legitimacy principle of knowledge" (39). If we follow Lyotard's logic, the crisis of scientific legitimacy implies a single objective and empirical truth, which allows for, in some sense, a multiplicity of truth. Fellow theorist of postmodernism Fredric Jameson (1991) suggested that "truth itself is part of the metaphysical baggage which poststructuralism seeks to abandon" (11). Jameson also discusses rejection of totality in postmodern thought, which "seems to suggest some privileged bird's-eye view of the whole is available, which is also the Truth" (331). In place of the totalizing view or perspective, within which a singular notion of objective reality ("Truth") was embedded, came instead a turn toward not just multiple truths but also partiality, and partial truths. Blaming postmodernists for post-truth is an act of blaming the messenger, as postmodernists named an epistemological shift as it was underway. Their work was an act of prediction, not of bringing this epistemological crisis into being.

Of course, we also find, in the work of French theorists Michel Foucault (1972) and Bruno Latour (1979), the notion of truth as a social production. From Foucault, we have gained the very valuable idea that what is named as true and becomes a "regime of truth" is necessarily embedded in power relations and does not necessarily mean or represent some kind of actual, underlying objective reality: "The problem does not consist in drawing the line between that in a discourse which falls under the category of scientificity or truth, and that which comes under some other category, but in seeing historically how effects of truth are produced within the discourses which in themselves are neither true nor false" (131, 119). For Foucault, the significance of scientific knowledge and accompanying institutions or "apparatuses"—in which he included academe, the military, and the media—was very much in policing and producing truths that facilitated a mode of governance and regulation of citizen-subjects (132). Yet post-truth society appears to represent a political shift insofar as the state appears to consolidate power notwithstanding open accusations of fabrication, deception, and lies. The state, in this sense, appears to be dedicated no longer to mobilizing a regime of truth but rather to subverting its necessity while the citizen-subject becomes governable without recourse to truth discourse.

In *Laboratory Life*, Bruno Latour (1979) theorizes how certain statements become facts by investigating researchers in a scientific lab: "While one set of agonistic forces pushes statement towards fact-like status, another set pushes it toward artefact-like status. . . . 'Reality' cannot be used to explain why a statement becomes a fact, since it is only after it has become a fact that the effect of reality is obtained" (180). The "circumstances of production" for fact are always "sociological and historical"; put differently, the production of objective "reality" is never seen as a totalizing ahistorical process but rather as socially embedded and contingent (105). A fact does not simply materialize but emerges on and through social production and work. Foucault and Latour, therefore, remind us that what becomes known and believed as true is never stable or tied to some kind of preexisting natural state. Thus, part of the history of the so-called post-truth was first understanding that "truth" is up for grabs, at least socially and historically, and not an absolute or pregiven realm of knowledge. Science, too, is not necessarily a truth discourse but rather as Gary Fine (2007) suggests: "Science is not a transparent window into truth. Rather, it is a *field of action*, used strategically to gain authority for assertion as about the material world" (58). However, science might still be *seen* as this transparent "window into truth" in popular culture. My way of understanding truth always interrogates and prioritizes the social actors' perception of truth and the material form of the obtained information over a notion of universal or objective truth.

This is precisely the line of thinking in Steven Shapin's (1994) *A Social History of Truth*, in which he investigates how social and historical conditions make particular agents, in this case, seventeenth-century gentlemen, emerge as reliable tellers of scientific truths. He argues, "The history of truth can be a social history because what we know about the world is arrived at, sustained, and recognized through collective action" (6). Truth-making is a social process as certain individuals become reliable speakers of accepted beliefs, and, according to Shapin, to trust is necessarily a moral process and act, "a moral bond between the individual and members of the community" (7). In this scholarly tradition of historicizing truth, Mary Poovey (1998) shows, in her history of British statisticians, that numerical data became seen as "immune from theory or interpretation," thus distinguishing a fact as "an epistemological unit" (14, 16). Poovey, like Shapin, insists that a series of historical and cultural factors shapes not only how knowledge is produced but also how it is organized and disseminated, as well as who can become a "knowledge producer" (16). Thus, the "conditions that make knowledge possible" undergird how subjects make sense of one another and the world (17).

Latour's (2010) thinking, in conversation with science studies on truth and facts (Pinch and Bijker 1987; Callicot 2015), on how certain kinds of information come to be recognized as factual continues in *On the Modern Cult of the Factish Gods*. Latour professes his interest in "the practical conditions of truth-telling" and, like Foucault, is more interested in *how* something becomes "true" and, therefore, does not proceed from a stance that truth is already existing in the world; they are not "primitive . . . primeval or the ground of mere perceptions" (114). Latour plays with the seeming epistemological difference between fact and fetish: "The word 'fact' seems to point to an external reality, and the word 'fetish' seems to designate the foolish beliefs of a subject" (21). How does one thing become known as true and objective while another as apparently backward and untrue? Latour suggests that both things require work and, furthermore, "conceal the intense work of construction" that results in this seeming opposition (21). The term "factish" is Latour's neologism that combines the two terms in order to call attention to and undermine this facile distinction between a constructed belief and an underlying reality. The "Moderns," for Latour, indulge in fantasies that "others" have fetishes, and they none; thus, maintaining a hierarchy between the Moderns' "anti-fetishist" worldview allows them to subject others' views as "belief" (30). I rely on Latour, among many applications of these logics, to theorize why CICAP protests superstition and employs probability to amend what to them are deeply nonmodern beliefs.

The concealment represents a significant way in which the fact can stand apart from the fetish. The factish or "fact-maker" is Latour's demarcation of a fact that gets to pass as "autonomous entities" through which the human fabrication remains obscured and hidden (35). Knowledge is thus opposed to belief in the following terms: to know something means that one acquires facts which are "real as long as [facts] are seen as *not made*" (63). On the other hand, belief is "powerful as long as [fetishes] are seen as not made" (63). Notice the difference between the *real* and the *powerful*, as the difference depends on religious notions of the invisible and incommensurate. Once again, the underlying work of social actors is hidden as the making of knowledge and belief remains metaphorically behind a curtain. If we consider *The Wizard of Oz*, the distinction might work as follows: an encounter with the projected face of the Great and Powerful Wizard would mean *knowing* him and would only become a religious "belief" in a godly wizard when we see the cranking and smoke of the machine producing the projection. The difference, as we will see, between the structure of knowledge matters: the televised news report versus the online blog.

With such cultural and socioeconomic conditions, the truth becomes an object of meta-analysis; that is, there was a historical moment in which citizens are influencing social theorists and social theorists are shaping citizens. Talking about truth and its manipulation became part of the larger cultural ethos. Regardless of the direction, the late 1990s and 2000s represented a kind of structure of feeling around truth and fact. In 2005, Stephen Colbert, famous for his work on the satirical news programs *The Daily Show* and *The Colbert Report*, coined the word "truthiness"—which he defined as "believing something that feels true, even if it isn't supported by fact"—and which was named Word of the Year by Merriam-Webster in 2006 (Rabin 2006; Weber 2016; Bierma 2006). With Donald Trump's presidency as an international focus, Oxford Dictionaries made "post-truth," meaning "relating to or denoting circumstances in which objective facts are less influential in shaping public opinion than appeals to emotion and personal belief" (Flood 2016), the 2016 word of the year. In 2016, Colbert used the term "Trumpiness" as the new truthiness: "Truthiness is believing something that feels true, even if it isn't supported by fact. . . . Truthiness has to feel true, but Trumpiness doesn't even have to do that" (McClennen 2016).[3]

The Rise of Mediatized Politics

The late twentieth century also represented the information revolution, the exponential rise of information production, sharing, and commodification in which the movement of information became synchronized with and fundamental to the movement of capital (Castells 1996). Kregg Hetherington (2017) argues that post-truth "retroactively" references the apparent "Truth" in post-Cold War governance, which centered around three key values: transparency to supposedly prevent state tyranny; information, particularly through the rise of the Internet; knowledge, especially the rise of the "knowledge economy," and Big Data.[4] The decades between Berlusconi in the 1990s and 2000s to the 2019 Five Star Movement are just as much shaped by the television heyday and Web 2.0 as they are other political and economic forces. Let me begin with Manuel Castells (2010) as he describes the contours of the communications, media, and sociopolitical revolution of the late twenty-first century amounting to "a qualitative change in human experience" (508). To him, the "network society" represents the new social structure and the "flows of messages and images between images" are fundamental to its production (508). Castells argues that the rise of informationalism "is oriented towards technological development, that is, toward the accumulation of knowledge and towards higher levels of complexity in information processing," and, in turn, "the expansion and rejuvenation of capitalism" (18, 19). At the same time, Castells argues that informationalism was fueled by globalization so information, like capital, became a "flow" and moved rapidly across borders, collapsing senses of time and space. The digital revolution, in which the spread of information circulates across digital means, has further intensified and reshaped "network society." In 2009, we had 1.5 billion Internet users and 3.4 billion wireless phone users (Castells 2010, xxv). In 2017, there are 4.1 billion Internet users and 4.93 billion wireless phone users ("Internet Users" 2017; "Number of Mobile" 2018).

It is important, therefore, to understand that the same decades in which philosophers began questioning truth as a monolithic whole coincided with the widest expansion of knowledge production and commodification that the earth has seen. While Castells focuses on how production of identity, the experience of virtual reality, and "timeless time" represent the most significant sociocultural implications of network society, I would argue that there was also, potentially, a kind of epistemological revelation wrought by these conditions. We must see that the late twentieth century was also witness to a mass volume of information and knowledge and unprecedented public access, which, arguably, enabled citizens to access competing forms of truth, thus potentially making more apparent that these competing discourses implied an instability of truth, a sense that knowledge was contingent on social production and dissemination. In other words, everyday actors, in being habituated to accessing networks of information, implicitly or experientially found that information was situated and multiple. The multiplicity and saturation of knowledge, in turn, could produce greater skepticism and doubt in relation to truth.

The new millennium saw the rise of media conglomerates, home computing, post-Fordist immaterial labor, and the birth of the Internet, offering unprecedented information and transforming the time and space of global connectivity, then World Wide Web 2.0, with the rise of social media. The Age of Information entails a series of massive shifts in how knowledge is produced, circulated, and consumed with dramatic consequences, as we are now only beginning to piece together, for governance and citizenship. Mediatization creates a consolidation of vast information, as it is often accompanied by a corporatization of the news media and thus a simplification of information into sellable consumer products. The mediatization of the political sphere has led to a reduction of political complexity into sellable and easily reproduced sound bites and regularly polarized political viewpoints, which appears to, in turn, favor candidates capable of self-branding and catering toward the memeable infrastructure of knowledge. Datafication is about the explosion of available information, made widely available across socioeconomic classes, as well the later iteration of this historical shift, in which data drawn from users became the new frontier for profit, surveillance, and consumption.

Yet the abundance of information has also slowly eroded how new knowledge is verified because citizens, on the one hand, have seemingly reliable resources with competing information, and, on the other hand, systematically confront the simulation of science or corporatization of science in which "studies" become the supposedly objective measure to promote everything from pharmaceutical products to gluten-free diets to face creams. Thus, the scientific method and the name of "real science" may be marshaled to support the discovery of Noah's Ark on creationist websites as much as they are deployed to verify climate change. Medical studies promoting the consumption of carbohydrates are promoted by the bread and flour industry just as Big Sugar is behind narratives that lack of exercise—as opposed to high calories of substances like sugar—is behind obesity. Dominic Boyer (2018) suggests that this marshaling of truth claims towards opposing facts is not the "absence of truth but rather the appearance of competing parallel spheres of veridiction in which ideological engines of truth-making radiate facts from normative institutional centers all the way into conspiratorial fringe speculation on both ends of the political spectrum" (85). Boyer's smartly observes that the erosion of facts does not necessarily mean the erosion of a desire to produce apparent facts.

Once citizen-consumers become aware that science, one "sphere of veridiction," is such a pliable discourse, the response may be skepticism, as their fundamental trust in science becomes eroded. However, certain social actors will begin to distinguish between science and pseudo- or corporatized science such that they are actually more deeply invested in the discursive superiority and perceived rationality of science (Tipaldo 2019). Put differently, the abundance of fake or faux science does not dissuade this group from seeing all science as bogus or bought. Instead, this group grows more vigilant—even hyperinvested and paranoid—about differentiating real science and true information. This is precisely the case I make for members of CICAP, the scientific skeptics who take to the streets to persuade fellow citizens to dismantle what they see as false beliefs in superstition and conspiracy theories. The risk of this epistemological position, however, is explored through the survivors of the L'Aquilan earthquake who equate their trust in scientific experts to a kind of anesthetization. I call this scientific anesthetization in which the trust in the superiority of scientific discourse overpowers other forms of sensory and historical information in ways detrimental to these social actors.

Jürgen Mittelstrass (2010) argues that the information age does not necessarily improve knowledge, but my point here is not about improving or not; rather, it is about exposing the underlying workings of knowledge production, which may yield a range of avowal or disavowal of information. Mittelstrass argues, "Knowledge and opinion become indistinguishable. . . . A niche for a new stupidity is opened" (20). To Mittelstrass, actors actually become less skeptical in engaging with information and thus become vulnerable to confusing information grounded in expertise as opposed to idiosyncrasy. Italian philosopher Maurizio Ferraris (2017) also frames the post-truth revolution in terms of the Internet revolution. While he recognizes that philosophers were theorizing truth, he jokes that Trump did not have to read Rorty or Foucault to enact post-truth, just as Cornelius Vanderbilt did not have to read Marx to enact capitalism. The "material premise" of this shift is what Ferraris calls "docu-media" (*documedialità*), which combines documentation and media, which, in turn, makes the Web an epicenter of both "absolute knowledge and non-knowledge." The docu-mediatized creates the underlying conditions—and consensus—that lack a coherent and stable shared reality: "The ideology that animates post-truth is the atomism of millions of people convinced they are right, not together, but alone. The center of this new ideology is the pretense to have the truth regardless of anything."[5] For Ferraris, the Web spurred a radical transition from a shared communal truth, which was reinforced and sustained through the Catholic Church in the past century, to a hyperindividualized collection of atomized realities, where the overlaps exist but a collectivity does not emerge.

Jean and John Comaroff (2000) are especially persuasive in their argument that neoliberal or "millennial" capitalism, in particular, what they call "occult capitalism," in which money seems to magically make wealth appear and disappear, engenders a turn to occult forms of accumulation and belief. Their work offers us another way to think about why late capitalism shapes belief and knowledge or erodes belief in uncontested facts and reality. Their theory, like Castells's, mines the historical and economic conditions of the late twentieth century and, in particular, the way in which capitalist regimes shape thinking and belief. In their understanding, a series of factors including the radical wealth disparity both within nations and globally, uncertainty, both economic and existential, and the commodification of the body, with a neoliberal capitalist regime, give rise to millennial thinking, a reenchantment of belief. The seemingly magical movement of capital gives rise to belief in and new practices of magic and occult, and the spread of religious movements as people make "appeals to the occult in pursuit of the secrets of capital" (313). The logics of capital exploit, plunder, and violently cleave global populations into widened groups of haves and have-nots. The unprecedented accumulation of capital incites enchantment as people attempt to make sense of its many unpredictable flows: "Magic is, everywhere, the science of the concrete, aimed at making sense of and acting upon the world—especially, but not only, among those who feel themselves disempowered, emasculated, disadvantaged" (Comaroff and Comaroff 2000, 319). Similarly, and referring to the 2008 market crash, Neal Curtis (2013) likens fundamentalist neoliberalism to a "delusional cult that even in crisis maintains its claims to truth" (74). What is new is not the ardent faith in neoliberal capitalism and market fundamentalism, but rather the way in which capitalism depends on or marshals knowledge production, evidence, and science (Curtis 2013, 74).

The tremendous and simultaneous revolution of information as a function of media, which produces a unique political and communicative shift, is also central to each of these theories (Castells 2010, 3). For Castells, the "real virtuality" of network society mutates time and produces both "simultaneity and timelessness" (419). If we consider Latour's notion that more obscuring in fact-production means greater supposed authenticity to facts, then we might consider that our time-warping media could only further mutate how consumers understand facts. On the one hand, facts might become even more robust because the process is further obscured, and thus the line between knowledge and belief thickens. Alternatively, one could also see how this temporal shift might instead expose fact-production mechanisms, thus rendering more facts into factishes, fact-like things but with visible gestation.

Jonathan Marshall et al. (2015) examine what they call disinformation society: "a realm in which devaluation of 'knowledge', failure of understanding, disorder of networks, disorder of property, disorder of financial institutions and continuous disruption of daily life is common, expected and arises within its successes" (2). The authors see capitalism and the commodification of information as a quintessential prerequisite to disinformation society: "restricting access to 'privileged' information, or providing inaccurate but saleable information, becomes a key tool for competitive advancement and profit" (98). The profitability of nontruths and partial truths fuels the cycle of disinformation and propagates its dissemination. Thus, they suggest that "relatively 'accurate' knowledge, especially public knowledge is devalued in favour of commodification, secrecy, pre-existing ideology" (247). They also suggest that the amount of information makes people unusually loyal to certain groups and sources of knowledge while hostile and suspicious, even "paranoid," toward other sources (250).[6]

So, we have two massive shifts: the intensification of global capitalism that renders bodies and beliefs more precarious and a mediatized saturation in information. The political sphere, in turn, mutates especially in relation to media (Roudakova 2017). The mediatization of politics means we might see more politicians "engulfed in a structural crisis of legitimacy," where political scandal feeds the twenty-four-hour news cycle and "personalized leadership, and increasingly isolated from the citizenry" becomes the norm (Castells 2010, 3). The "hyperritualized magic" of the state works through and in the media, and social actors necessarily have multiplied their interactions as governance becomes fragmented through the array of information and news, accessed and resonated across multiple forms and genres (Comaroff and Comaroff 2000, 329; Kennedy 2017).[7]

Disinformation as "Fake News"

In October 2017, Laura Boldrini, the president of the Italian lower house of Parliament, introduced a plan to help Italian children discern real news from fake news (Horowitz 2017). Instruction would begin in October 2017 for eight thousand high schools in Italy, and improve students' reading literacy of digital media. Boldrini said, "Fake news drips drops of poison into our daily web diet and we end up infected without even realizing it." Pope Francis dedicated his 2018 World Communications Day address to fake news (Francis 2018). Pope Francis positions fake news, "the capacity to twist the truth," as part of the human condition, deriving from "the earliest times" with Cain and Abel and the Tower of Babel. He warns citizens that such disinformation requires knowledge to discern, as it "is often based on deliberately evasive and subtly misleading rhetoric and at times the use of sophisticated psychological mechanisms" (Francis 2018). Rather, the kind of literacy practice one needs is a "profound and careful process of discernment," which recognizes how the capacity for reading is a learned literacy practice. He ends his statement with a call for a "journalism of peace," which is committed to truth and works actively against the dissemination of falsehoods. On the one hand, true/false appears to be a binary, yet in practice, there is a range of how truths live on a spectrum, ranging from subtle "misleading rhetoric," as Pope Francis suggests, to active fabrication and deliberate deception.

In other words, the notion of "fake news," specifically, and post-truth, more broadly, does not suggest an after or beyond the truth, but rather a kind of subservient position of truth to fabrication. It is also helpful to conceptualize the shift as a para-truth as "para" designates "at or to one side of" and "pertaining to or occupying two positions," a mix of beside, near, by, beyond, alongside, and contrary. The prefix came to "designate objects or activities auxiliary to or derivative of that denoted by the base word and hence abnormal or defective" ("Para" 2017). Consider the term "paranormal." Paranormal is not "after" or sequential to "normal"; it combines a sense of being both beside and beyond normal. Whereas the prefix "super" (supernormal, supernatural) is both "above" and beyond, the prefix "para" leaves the temporal progression and moral hierarchy far more ambiguous. In 2000, anthropologist George E. Marcus published *Para-Sites* and highlights how "para" emphasizes the notion of "sideness" but also "amiss, faulty, irregular, disordered, improper, wrong. Also expressing subsidiary relation, alteration, perversion, simulation" (6). Similarly, Victor E. Taylor's (2000) *Para/Inquiry: Postmodern Religion and Culture* also theorizes "para" as a postmodern phenomenon and in particular how "para" suggests an "unstable" event in relation to the thinking and knowing of a particular context (16).

The political undermining is far closer to normalizing defective truths or paratruths; moreover, citizens are able to process multiple forms of information; the "alternative facts" are, in essence, "at one side of" the facts. Both are available, but what is new is how this kind of millennial political regime engenders a robust stickiness of an inferior or suspect version of reality. Enchantment and the occult may be desirable, but they are within a long Western tradition in which scientific knowledge is imagined as a superior truth: empirical, universal, objective, rational. Yet this notion of "truth," which represents a supposedly single unitary reality in the realm of science, competes with Italian cultural spectacle in politics, media, and day-to-day life where the almost-, quasi-, or pseudotrue dominates.

The meaning of a truth society, then, is an assemblage of forms, affect, materialities, and knowledge that create the conditions in which in semitruths or "near-realities" emerge "alongside" reality and become more accepted, more apparently rational, and more plausible. In a society like millennial Italy, we find both enchantment and scientific rationalism. To suggest that there is a battle between reenchantment and scientific rationalism is to miss the deeper epistemological problem. What citizens have witnessed is a crisis in how facts are authenticated, how certain propositions come to be verified as true and false, and how the multiplicity of propositions leaves that process open for political manipulation and exploitation. Another effect of mediatized politics is the rise in rationalism and hyperinvestment in science. I track these patterns in the work of CICAP's scientific skeptics and the trial against earthquake scientists in L'Aquila. The investigation of the trial allows me to investigate and explore another potential consequence of truth societies: the higher stakes of scientific knowledge. Both CICAP's mobilization and the L'Aquilan legal case represent cases in which we see how Italy's culture of artifice and mediatized disinformation, in some sense, backfires and people overinvest and even fetishize what is apparently rational and objective truth and, in turn, science and scientists.

In defining "para-site," Marcus also has witty definitions of "parasite," which we might consider compelling for forms of politics, calling them "the wily transgressor within" and "clever, self-involved exploiters without redeeming social value" (7).[8] Parasites imply an inside job. Para-sitic truth politics confuses the terrain in that the host becomes an ecosystem for the successful operation of the parasites. If we extend these logics to the notion of truth societies, in which the prevalent practice undermines objective reality by populating the knowledge sphere with semi- and false truths, then it is as if the new understandings of the world become like parasites and redesign the circuity of cultural knowledge. We must excavate the cultural and historically specific ways that political forms shape and reshape understandings of the world. The affective dimension of the semitruth is at least some degree of wiliness, as if it is a secret ruse, which reminds me of the Italian notion of cleverness (*furbo*). The term *furbo* lies at the nexus of smart, clandestine, and unscrupulous. Berlusconi's cleverness involved a unique manipulation in that the outrageous artifice was presented as so outright and plain that it made him appear more exposed, and thus, knowable. Put differently, Berlusconi's outlandish and politically rogue statements were so manifestly absurd that he appeared to expose his real or true self to the Italian constituency. Thus, one clever political strategy of Berlusconi, and later Beppe Grillo, is how the fabrication can appear as genuine; the trick appears so overt that the covert obfuscation seems to disappear.

Still, the notion of truth or paratruth alone seems to evade the agents of truth claims, who become socially heard as speaking a reliable truth or valuable proposition about the world. As we will see, I find that trustworthiness is not exclusively gained by speaking about or even seeming to speak truth, as we find in the case of Berlusconi. Rather, the position of a truth-teller is viewed cynically, not necessarily held in esteem or trusted, and often under attack, as we find with news media and the emergence of "fake news" as a category. However, in the case of L'Aquila and the scientists on trial for what amounted to deadly public reassurances, the question is about what historical

Figure 0.2. Venice graffiti, June 2017. Photo by author.

and social conditions made citizens trust in what the scientists said. How does someone become authorized to speak reliably about the natural world?

The question was posed by Steven Shapin (1994) on why seventeenth-century English gentlemen became "truth-tellers" (65). Shapin analyzes values such as "honor and gentility," and how the status of being a gentleman allowed for and maintained social order. I adopt Shapin's understanding that trust in certain truth-tellers represents a moral practice and "the identification of trustworthy agents is necessary to the constitution of any body of knowledge" (xxvi). In *The Scientific Life*, Shapin (2014) terms them "truth-speakers" and highlights our need to identify that we track "what kinds of people, and what kind of attributed and acted-upon characteristics, are the bearers of our most potent forms of knowledge" (6). Dominic Boyer (2018) has argued that the erosion of trust in media and democratic processes represent "symptoms of the anxious and fugitive character of liberal expertise" (85). I also track when the trust and expertise break down and how the emergence of untrustworthy experts gives us insight into a culturally and historically specific body of knowledge.[9] Here, too, then, we might consider why CICAP promotes the distrust of conspiracy, superstition, and homeopathic medicine.

Silvio Berlusconi and Italy as the World's Laboratory

When Enrico Letta was elected the prime minister of Italy in the spring of 2013, he told the Italian public he would speak in "the subversive language of truth" (ANSA 2013). His election came months after February's suspenseful deadlock between former prime minister Silvio Berlusconi, who had been sentenced to seven years in prison and banned from Italian politics for his involvement in underage prostitution and abusing his power, and comedian-cum-populist Beppe Grillo and his Five Star Movement, and leftist candidate PierLuigi Bersani (Davies 2013). Meanwhile the Italian economy spurred fears of its becoming "another Greece": the *Economist* (2013) reported anxiety and concern for Italian banks and an increase in bad loans and high national debt; the *New York Times*, in an article titled "On the Brink," observed Italy's economy was growing at half the rate of the euro currency union's economy and facing "one of the worst recessions of any euro zone country," with unemployment rates at 11.7 percent and youth unemployment at nearly 40 percent (Alderman 2013).

The 2013 election, and Letta's single year as prime minister, represented insight into the Berlusconi years as we find economic crisis and the emergence of the new populist movement with a comedian, Grillo, at the helm. To some scholars, the 2013 election revealed "that Italians have perceived the fiasco of all the political programs that have been tried since 1994" (Orsina 2014, 135). Media mogul Silvio Berlusconi entered Italian politics in 1994 with his right-wing political party, *Forza Italia*, fashioning himself as a successful businessman and external to Italian politics as usual. Before Berlusconi's reelection in 2001, Berlusconi's single year in office was succeeded by Lamberto Dini, Romano Prodi, Massimo D'Alema, and Giuliano Amato. Berlusconi then held office between 2001 and 2011, with just one two-year turnaround to Democratic Party leader Romano Prodi between 2006 and 2008.

In addition to Berlusconi's shaping economic policy by stripping workplace safeguards and rendering the economy more precarious, Berlusconi's years in Italian government were marked by one legal or high-profile salacious scandal after another: tax evasion, underage prostitution, hosting sex parties, and corruption (Ginsborg 2005; Allum 2011). Berlusconi resigned in 2011 but, importantly, had not only held the longest-running term as prime minister in Italian history but also held the national spotlight for over twenty years.[10] After being found guilty for soliciting an underage prostitute and for abuses of power, Berlusconi was expelled by the Senate in November 2013 and barred from running from political office for two years (Orsina 2014, 144). However, he staged a comeback in 2017 and served as the figurehead of his Forza Italia party in the 2018 elections, even though he could not yet legally serve as prime minister. Grillo's Five Star Movement and the right-wing Northern League (Lega Nord) parties gained a majority, but no single party garnered a sufficient majority to rule; thus, Five Star and Northern League held a ruling coalition from June 2018 to September 2019. After a no-confidence vote in September 2019, Giuseppe Conte remained prime minister with a new political coalition including Partito Democratico and the Five Star Movement.

Why, then, would Letta distinguish himself as a truth-teller? The easy answer is Berlusconi's embroilment in scandal around his resignation in 2011 and, building up to that year, scandals involving lies and denial and "misunderstandings" for Berlusconi. Yet Berlusconi had also long been known for gaffes and off-putting humor, which sent a favorable message to Italians that "politics was not so serious really and people in government were not in any way superior to ordinary people" (Orsina 2014, 73). Berlusconism relies on a kind of theatrics and humor, his own version of reality, especially in defending himself against a constant siege of scandal. Let us remember that Berlusconi was a media mogul who owned a massive media conglomeration, which was also instrumental in rapidly passing sweeping neoliberal economic policy in the 1990s and 2000s. In fact, Berlusconi's Mediaset company controlled 90 percent of the Italian television market during his 2001–6 administration as well as a company, Publitalian, that controlled 60 percent of television advertising (Benini 2012, 88).

Many of the Western political forms we have seen in the past two decades in Italy are not unique to Italy but have regularly emerged sooner than in other liberal democracies, especially Berlusconi's control of television media. Consider Silvio Berlusconi as an innovator in now globally circulating political forms: the appeal of antiestablishment, anti-intellectual "outsider" political candidates, a strong neoliberal economic and policy agenda, the deeper entrenchment of cronyistic power and patriarchal norms.[11] But Berlusconism was possible only because of a much more profound change in Western epistemological groundwork which included the *televised* mediatization of politics, in which politics became visually consumed on TV and twenty-four-hour news consolidated sound bites of politicians' speech, the ease of disseminating disinformation on the Internet, and in new media channels, most broadly, the onslaught of information in the digital age and the fraught process of verifying truths or postfact society (Boyer 2013a).

We might see Berlusconi as an innovator in post-truth politics where affective engagement and belief override objective truth but also because of his control over the material dissemination of information: television. Even if I critique the implications of this form of power, I recognize Berlusconi's prescience in terms of the velocity in which he ushered in economic change and especially in mounting a media-dependent theatrical governing regime. He commanded the material infrastructure of information, television, which in the 1990s and early 2000s was the leading form of consumer media in Italy. In turn, truth-production in Italy and understandings of scientific objectivity also came into crisis, unfolding in rapid and intensified ways. Valentina Fulginiti (2016) analyzes the fiction and nonfiction literature of Berlusconi that results in the "fictionalization of politics," which subverts the truth by focusing on storytelling and mediatization (111).[12] In discussing post-truth, philosopher Maurizio Ferraris (2017) said in knowing Italian "mediatized populism" for the past twenty years, he's earned the reputation for being "prophetic" from his American colleagues. While truthiness and the Bush administration help ground this epistemological shift in the United States, in Italy, we can find that Silvio Berlusconi, media mogul who dominated Italian politics for over two decades, had already masterminded the art of mediatized truth-bending politics.[13] Italians have told me many times over the years and scholars agree that Italy is like the world's laboratory, a place that germinates global trends (Benini 2012 88).

On a deeper level, Berlusconi's political creation that, yes, arguably predated other similar political eruptions was not only a mostly illiberal and right-wing xenophobic populism but a political regime not just tied to lies and spectacle, but one in which objective reality was called into question. Berlusconi has been famous for contesting political judgments, legal cases, and most broadly, criticism against him by suggesting he was "misunderstood" (*frainteso*) (Newman 2017); these consistent disputes with reality, in turn, eroded trust and faith of an already cynical population. In late twentieth-century and early twenty-first-century Italy, we see the collision of sociohistorical forces: descaffolding of the welfare state, erosion of steady and reliable labor, and drastic shrinking of the middle class, and unprecedented access to information and the Internet revolution. Thus, in this investigation of Berlusconi as a generator—and, to some extent, original tinkerer—of a post-truth political regime, there are lessons for other nations and histories, including but also well beyond the United States.[14]

Epistopolitics

Attending to Italy's political phenomena—from theatrical and fact-bending politics, Internet populism, and algorithmic democracy and the emergency governance—requires that we map both their epistemological precursors and aftermath, such as the televised mediatization of news, antiscience discourse, and the fetishization of scientific discourse. In this sense, then, my ethnographic objects are the material tracks of these profound epistemological shifts. What kind of ways of knowing and processes of accumulating knowledge set the stage for these political forms? How do these new political arrangements shape not only how we know the world but what we think is knowable? The epistemological forms of the digital age have, in some ways, been overlooked as a significant contributor to political innovation in late modernity. The material and structural fallout of new knowledge practices and arrangements—from network news to algorithms—represents my central ethnographic objects.

The trial in L'Aquila is also a prime result of this form of Berlusconism and postfact society. The trial contains several additional political forms of great importance: the question of causality and culpability for natural disaster, the legal accountability of scientists' news briefings, and crisis governance that relies on legal exception and subversion of due process. Moreover, the epistemological undercurrents that shape L'Aquila include disinformation and fake news and the simultaneous skepticism toward and hyperinvestment in scientific thought, and perhaps least obviously the Anthropocene, the notion of human-created climate change which represents a paradigmatic shift in how we know and understand the earth and what we term natural disaster, and conspiracy theories, which are enormously facilitated by online reading practices and forms.

Beppe Grillo and the Five Star Movement revolution represent a significant political revolution in the form of right-wing antiestablishment populism, the discourse of enchantment and magic, antiscience sentiment marshaled as political platform, and algorithmic democracy through the party's famous digital platform, Rousseau. Some have argued that the rise of populism stems from emotional responses to leaders and harkens back to an enlightenment division between emotion and reason (Davies 2019). Such analysis misses the deeper shift in the age of information, and how information processing, algorithm society, and media shape the rise of populism. Berlusconi, the man with the majority control of Italian television, gets surpassed by Grillo, a man whose blog became the number one Internet site in Italy and, globally, in the top ten blogs (Loucaides 2019, 87). This is not an incidental part of the story; it *is* the story. The massive move from right-wing, patriarchal establishment Berlusconism to Internet-based populism was fundamentally a product of the television-to-Internet shift. The glowing screens with fixed daily programming in Italian family kitchens at dinnertime became individual citizens reading online texts in their bedrooms. Head shots of Berlusconi on talk shows, the camera closing in on his made-up face and cosmetic surgery, gave way to YouTube videos of Grillo, live in front of thousands. National news, censored and scanned by Berlusconi-owned Mediaset, became Grillo's private blogs offering their own counternews, referencing other websites as supporting evidence.

Grillo and the Five Star Movement's rise may be understood as an effect of and a response to this technological shift. The mediatization of politics did not change but its form did, from the television cable networks to algorithms and servers. Just as we need to understand television channels and advertising to understand Berlusconi, so too must we examine algorithmic organization of information online to best comprehend Grillo and the Five Star Movement. At each turn, politics always was connected to how knowledge and information were structured, circulated, and processed.

If we invest analytically in unearthing these patterns, then Italy's late millennial world clearly shows us the locked embrace of how we know and how we govern, what we assume to be real and what we assume will happen politically, who are political constituents and who are knowledge creators, who makes truth claims and who makes authoritative decisions, and what secretly undergirds knowledge and what secretly undergirds power. In all of these examples, the "embrace" may be cause and effect, it may be parallel empowerment, or it may be inversely related. For example, consider modes of knowing and governance. Media consumption articulates a particular practice of gaining information that tends to be highly reproducible, coordinated among various invested actors and economically aligned resources, and contains simplified if not outright reductive messages. The practice of mediatization has a direct effect on how leaders govern certainly insofar as media-savvy and image-conscious politicians like Berlusconi and Grillo are concerned (Hajek and Salerno 2014). The question of who makes truth claims and who makes authoritative decisions may have once been the same group of political leaders. Yet in the postfact political world such as in L'Aquila during and after the 2009 earthquake, the scientists were authorized as truth-makers in their reassurances to citizens that no earthquake would come after a series of swarms or smaller quakes. But the key decision-makers were members of local governance hoping to use the news conference to quell rising public panic and later who suspended democratic governance in their authoritarian control of the post-earthquake housing camps. The chaos of postdisaster L'Aquila together with Berlusconi's charismatic promise of renewal—not a truth discourse about health or safety—served as the central means to erect a political authoritarian rule by exception, not law, which may be read as part of global trend towards increasingly widespread authoritarianism (Boyer 2018).

Just as the age of the Internet has led to a seemingly commonplace decoupling of content and form, new political forms rely on a decoupling of the content and form of knowledge. Berlusconi's misogynistic remark about Angela Merkel became viral. The quickly reproducible remark was decontextualized on television and online in ways that stripped the gaffe of its patriarchal gravity. Therefore, a viral politician's joke was a new form that obscured the content of misogyny and ageism. Infamous earthquake predictor Giuliano Giuliani garnered celebrity status as he predicted earthquakes based on scientifically antiquated signals such as radon emissions. Here L'Aquila's "Cassandra's" prediction was made possible precisely because of the rapid fire and easy dissemination of Giuliani's thinking via online channels and in news media. The Five Star Movement's deputy prime minister Luigi Di Maio seems so perfectly suited to constituent preferences that he's been accused of being fabricated from algorithmic calculations (Minuz 2018). Di Maio may be one of the first political figures whose personality and appearance—his "content"—seem separable from his "form"—how he came to be, the algorithmic forms of knowledge undergirding his popularity. Thus, at every turn, epistemological shifts of the new millennium have and produce new material circumstances and political forms, which in turn create and limit ways of knowing.

Overview of Chapters

The book moves from establishing television and print media as foundational to Berlusconi's antiestablishment establishment politics to its many aftermaths. The immediate aftermath is about how science became seen as worth saving and a rallying cry to mobilize to the streets. The more consequential outcome was the rise of Beppe Grillo and the Five Star Movement, where we will find the Internet and algorithmic processes form a new kind of digital populism. Next, in L'Aquila, we examine why Italy was one of the first countries to attempt to hold disinformation and "fake news" legally accountable, and how social actors hyperinvest, at a cost, in scientific thinking. The conclusion posits how the current hegemonic form of knowledge, customized Internet knowledge, shapes and will shape forms of governance and politics.

Chapter 1, "Manifest Disguise and Mediatized Politics," circles around a mystery of late twentieth-century Italy: despite widely shared agreement on his corruption and ineptitude, Silvio Berlusconi enchanted citizens and dominated Italian politics for nearly twenty years. How does a dazzling political culture change how Italians discerned what counted as accurate and reliable information, and which actors might be trusted to offer the facts? To answer these questions, I turn to one of the world's first fake news programs, *Striscia la Notizia* (The news is creeping) and its plush mascot Gabibbo, a human-sized red puppet who is praised as a civil defender. Gabibbo helps me unravel why Italy become a site in which puppets talking politics were more reliable than puppet-like politicians. I suggest that postwar political spectacle gave way to a widespread popular cynicism capable of simultaneously propelling former prime minister Silvio Berlusconi's peculiar popularity and puppets seen as truth-tellers. I make sense of why Italian politics and politicians were thrust toward mediatized theatrical politics—where the form of politics upstaged the content of political discourse—sooner and more dramatically than in other Western democracies. An ethos of cynicism, together with the intensification of theatrical and mediatized politics, I argue, has made unhidden artifice—one in which spectacle is fully undisguised and overt—appealing and affective in Italy. I show how Italian citizens have become habituated to the carefully staged exhibitionism of national politics and how, in turn, political spectacle led to belief in and fascination with the supernatural and magical but also to a desire for reliable truth. Berlusconian Italy became an age of artifice and populist political enchantment with significant effects for the production of knowledge and what counts as truth.

In chapter 2, "The Soldiers of Rationality," I examine the yearly demonstration organized by the Italian Committee for the Investigation of Pseudoscientific Claims (CICAP), "The Day against Superstition," in which the group stages a national protest against superstitious belief such as the belief in spooky cats, unlucky numbers, and the avoidance of mirror breaking. Members identify as "soldiers of rationality" who protect Italy's credulous public, whom they see as victims of dangerously irrational and misguided beliefs. The group promotes scientific knowledge and debunks apparently irrational convictions, which enables them to lump together witchcraft, 9/11 conspiracy theories, UFOs, and the Shroud of Turin. To them, any form of experience that relies on faith rather than scientific measurement is erroneous, even as their activism occasionally becomes a proselytization of science. What is at stake for these self-proclaimed "skeptics" and "soldiers of rationality" in these nonscientific beliefs? The chapter unravels the mystery of why this group has become suddenly urgent, especially in contemporary Italy, a country known for its centuries-old belief in witchcraft, magic, and superstition. The crisis, I propose, that led to the rise and intensity of Italy's scientific skeptics was not only economic and political but also epistemological: a crisis of how facts are verified and authenticated. Pro-science activism, therefore, becomes an occasion to perform new practices and rituals of rationality and science in the context of media artifice and political disinformation. CICAP members insist on grounding truth in quantifiable and valid claims, in contradistinction to the highly malleable fictions of Berlusconian Italy. In this sense, the insistence of the "soldiers of rationality" on the extreme superiority of scientific knowledge might be seen as reactionary, protesting the seemingly dissolving boundary between fact and fiction.

Chapter 3, titled "The Rise of Algorithm Populism," scrutinizes the political career of comedian-cum-politician Beppe Grillo, who founded the party that currently co-rules Italy, the so-called algorithm party: the Five Star Movement (Movimento Cinque Stelle). How did Italy shift from the glossy prepackaged world of Berlusconi toward a grassroots, Internet-driven, and algorithmic political movement? The Five Star Movement has been called a protest movement, populism, antipolitics, and even anarchism; certainly, Grillo's flair for humor and satire challenges the Berlusconian status quo. Grillo also deploys supernatural humor, and political suspicion deserves more credit for propelling his movement than has otherwise been recognized. His otherworldly humor and cynical hypotheses do not just speak to brewing cultural and economic anxieties, but rather represent a kind of wink to citizens. He, too, recognizes the profound and ghastly irrationality of contemporary Italian politics. Though Grillo's theories might offer citizens a scapegoat or an alternative explanation for Italy's socioeconomic crises, the labor of the supernatural, I think, is actually epistemological: Grillo's ghost stories are really about how supposed facts are made, processed, and obstructed. For Grillo, Italy is haunted by a mass-produced false reality, which is propagated through news media and disinformation. Correcting how Italians process information and making them skeptical of televised media in favor of algorithmic programs and online data became, for Grillo, a high-stakes political result of ushering Italy's first populist party into the seat of government.

In chapter 4, "The Trial against Disinformation," I consider the trial and subsequent conviction of scientists following the April 2009 earthquake in the city of L'Aquila. In 2012, members of the National Commission for the

Forecast and Prevention of Major Risks (Commissione Grandi Rischi) were charged with manslaughter and grievous injury for issuing public reassurances, which were purported to frame victims' choice to stay home and, subsequently, led to their deaths. I show how the L'Aquila trial represents a seismic rupture not just of land, but of belief; the trial became a judicial experiment in holding disinformation accountable. I explore why citizens of L'Aquila might have found scientists' public statements credible. For some, those statements momentarily appeared as absolute as supernatural forces and "anesthetized" them. The incrimination of the Risk Commission, in other words, meant that Italians had faith in the moral capabilities of politicians and the purity of scientific truth, suspending their more rational skepticism to the contrary. At the same time, the incrimination of L'Aquila scientists represented, at least in part, a kind of misplaced indictment of Italian media. Put differently, I argue the court of L'Aquila held media utterances issued by scientists accountable for human tragedy, which indirectly summons the Italian media's propensity for deception and lies. Based on historical, legal, and ethnographic research, this chapter makes sense of why certain scientists in certain places are held accountable for a natural disaster's human casualties. I suggest that the trial allowed the Italian courts to dabble in the metaphysics of culpability—important for a country in which the absolution of guilt is the quintessential tenet of Catholicism. How do the courts create new causality narratives in order to make sense of unexpected death and misfortune? Thus, and for not the first time in Italian or world history, the court serves as central mediator between science, government, and the production of knowledge.

In chapter 5, "Scientific Anesthetization in the Anthropocene," I rethink the L'Aquila trial in a larger, socioecological context in which natural disaster has been reimagined as radically altered by human action, or the Anthropocene, asking how this paradigmatic shift in understanding global ecology shapes how human agency and accountability are viewed in a single disaster. For the first time ever, society has reconceptualized natural disasters as resulting, at least in part, from human action, not natural force. It is not surprising, then, that with this age come new fantasies about predicting, controlling, and managing human disaster. If human action can shape massive geologic forces, the logic goes, then surely it can be as potent to stop, correct, and manage them. In this sense, I reevaluate the L'Aquila trial as revealing a newly emerging set of geological cosmologies. Perhaps the scientists were held accountable because of a growing—and sometimes magical—belief that the earth, and nature as a whole, could be destroyed but also saved by humans and sophisticated scientific knowledge. Delving into interviews and tracking online discourses, I track millennial conspiracies about man-made natural disasters in Italy, which include giant warming machines that cause earthquakes and "chemtrails," toxic airborne emissions deliberately disseminated via aircraft. I argue that these growing rumors and conspiracy theories about deliberate and intentional disaster, which may initially sound supernatural, may, in fact, offer a savvy critique of human complicity in natural disaster. This chapter explores how both L'Aquila's trial and geological conspiracy theories are tethered to an Anthropocenic vision of the world, as well as to shifting moral understandings of God and the earth.

In the conclusion, "The Mirrored Window Society" I return to the emergence of the Five Star Movement and its dependence on algorithmic technologies. From digital and government surveillance to Wall Street, the locus of knowing the world and knowledge of the self, once more firmly and singularly rooted in the human corporeal self, has been shifted toward mathematical algorithms and computers. How does living in a world in which knowledge is not just run through algorithms but also customized to the individual shape what we think about the world and who we think should rule it? In an age dominated by highly sophisticated but indecipherable and invisible forms of intelligence, I posit what these new material infrastructures of knowledge mean for democratic governance.

Chapter 1

Manifest Disguise and Mediatized Politics

There is no one on the world stage who can compete with me.

—Silvio Berlusconi

A head of state that tells jokes everywhere . . . a [head of state] who would be merely a blemish if he weren't a national tragedy.

—Eugenio Scalfari, "Lo Stato disossato e i pasticci elettorali"

Before his resignation in November 2011, Silvio Berlusconi was Italy's second-longest-serving prime minister; he is the world's thirty-seventh wealthiest man, a media mogul who owned the majority of Italian television networks and television advertising, and leader of the center-right political party Il popolo della Libertà (Bigi et al. 2011, 151; Benini 2012, 88; Edwards 2005; Ginsborg 2005).[1] His charismatic though scandal-ridden leadership has been a defining part of contemporary Italy since his first government mandate in 1994, so much so that historian David Gilmour has described him as having "a hold over the Italian people as no other politician . . . since Mussolini" (2011, 380). According to Slavoj Žižek (2009), he manages to fuse "permissive-liberal technocratism and fundamentalist populism."[2] Nicknamed "the Knight" (Il Cavaliere), he has further consolidated his "playboy" image with plastic surgery, hair implants, and candid discussion of his sex drive (Allum 2011, 289).[3] For most of his three terms in office, he was the target of judicial proceedings and accusations of criminal misdeeds ranging from tax fraud to corruption to Mafia involvement, and he was infamous for sex-related scandals involving marital infidelity, divorce, and his organization of sex parties with underage prostitutes. On that note, the creator and producer of Italy's most popular news parody show, *Striscia la Notizia* (The news is creeping; hereafter, *Striscia*), had a resonant one-liner for Berlusconi's sex scandals: "Everyone knows those girls got paid just to listen to his horrible jokes" (Cazzullo 2011).

In addition to his spectacular improprieties, Berlusconi is known for his hilariously awful remarks and public gaffes. Let us review some of his most audacious: calling President Barack Obama "suntanned" (*abbronzato*; Merlo 2011); suggesting the political party Social Republic (Repubblica Sociale) rename itself "Go Pussy" (Forza Gnocca); gesturing "gimme five" to former German chancellor Gerhard Schroeder (Messina 2002); telling a German member of the European Parliament that he would be a perfect cinematic Nazi commander (Billig 2005, 177); proclaiming German chancellor Angela Merkel an "unscrewable lardass" (*la culona inchiavabile*; Telese 2011); uttering to the then president of the European Union Anders Fogh Rasmussen that Rasmussen was "the most beautiful premier in Europe" (Telese 2011); and even quipping, "Mussolini never killed anyone. [He] used to send people on vacation in internal exile" (Allum 2011, 289).

Though Berlusconi has called his style "smile diplomacy" (*la diplomazia del sorriso*), such comments have gotten him dubbed "Sir Gaffe" (Signore delle Gaffe) and "Al Boorish" (Al Cafone, a play on Al Capone; Messina 2002). And to his critics, he has responded with more jokes, saying, "The communists [have] no sense of humor" (Gilmour 2011, 381).[4] Whether people find him funny is not my interest. I start from the fact that Berlusconi has crafted himself as humorous and relatable and, despite wide agreement on his ineptitude and moral bankruptcy, has captured broad political support in Italy since 1994. His cumulative years in office make him the longest-serving postwar prime minister.

How can humor, particularly humor made into tidy televised tidbits, serve a political leader who is widely seen as inept? What are the distinguishing features of political satire when a country's own prime minister is both majority owner of Italian television and the object of ridicule? We might link Italy's news parody *Striscia* to the rise across many late-liberal Western democracies of "overidentification" humor, which so genuinely mimics the normative form of discourse that it becomes hard to tell whether it is ridicule or support (Boyer and Yurchak 2010, 181), as well as to "theatrical political

Figure 1.1. Italian prime minister Silvio Berlusconi in Milan in April 2010. Photo by Bruno Cordioli.

activism" (Haugerud 2012, 153). Political humor and satire, particularly in the West, have become more challenging to decipher as polarized political messages are increasingly synchronized and "mediatized" as well as digital (Campus 2010; Boyer 2013a). Viewers, in other words, need to do more imaginative work to figure out that a lavishly dressed and presumably wealthy citizen announcing his or her ardent support of George W. Bush's tax cuts (Haugerud 2012) or that Stephen Colbert's testimony to Congress on migrant labor (Boyer and Yurchak 2011) is actually undermining, not endorsing, the position being advocated. Writing on irony, James W. Fernandez and Mary Taylor Huber make the observation that irony can "afford political circumstances where direct dissent is hard to formulate, risky, or unwise" (2001, 5). Pathways to direct dissent may be limited or obfuscated in a variety of ways, through direct censorship of and constraints on media but also through media's own activities, such that increasingly homogenized political messages are engineered for the public (Baym and Jones 2012; Boyer and Yurchak 2010). Such circumstances fittingly describe Italy in the 1990s and thus inform my analysis of *Striscia*. However, neither *Striscia* nor Berlusconi's own humor can be complexly characterized without a deeper examination of why televised spectacle and satire can be appealing to an Italian citizenry and politically strategic for a head of state. Examining *Striscia* becomes a way to consider how the material ways in which citizens encountered news media and information became the foundation to Berlusconi's unique right-wing, patriarchal mediatized political regime.

Born to a middle-class Milanese family, Berlusconi became a successful entrepreneur who, by 1978, began Fininvest, a holding company of several media channels. Events in 1989 remapped Italian politics: the Christian Democrats (Democrazia Cristiana, DC) lost stability, finally disbanding in 1994, and the Communist Party (Partito Comunista Italiano) was dwindling, morphing into the less significant Democratic Party of the Left (Partito Democratico della Sinistra, PDS) in 1991 (Allum 2011, 285); the late 1980s marked the end of Italy's so-called First Republic (1946–92) and signified a slowly loosening hold of Catholicism and Marxism in national discourse (Ginsborg 2005).[5] This was the crucial moment in which Berlusconi's new party, Forza Italia (Go Italy, FI), could ally with the right-wing former fascist party National Alliance (Alleanza Nazionale, AN) and a separatist party, the Northern League (Lega Nord, LN), and win the 1994 election. By the 2001 elections, Berlusconi had been in and out of office twice, succeeded by four different short-term left-wing leaders; for his reelection campaign in 2001, he distributed his life story, *Una storia italiana* (An Italian story), to Italian households, presenting himself as a "commonplace man whose ordinary virtues . . . had made him enormously rich" (Edwards 2005, 226). He returned to office in 2001, and his leadership endured until 2011, interrupted only by the two-year interim of the Olive Tree coalition's Romano Prodi, a Democrat (May 2006–May 2008). In 2018, he ran as the figurehead of Forza Italia, which did not gain the majority in the March elections. After his ban from Italian politics ended in 2019, he was elected and has served as a member of the European Parliament since July 2019.

Many Italians have been deeply suspicious about how Berlusconi has used the judiciary to favor his business interests. In 2001 he loosened laws on false accounting and reset a statute of limitations to annul several suits against his own company. A 2003 law he enacted saved his television channel Rete 4, and a 2004 law limited media ownership but left his own company, Mediaset, alone (Edwards 2005, 236–237). After his 2008 reelection, he was still embroiled in legal troubles for his involvement with underage prostitutes and more legal maneuvering around his own prosecution. He also had waning popular support. He survived a December 2010 no-confidence motion by only three votes, and he resigned eleven months later (Allum 2011, 281). It is not uncommon to call Berlusconi's political career an "anomaly" (Edwards 2005, 236). Erik Jones (2009, 39–40) uses "anomaly" because of the rare political circumstances that led to Berlusconi's rise to power: the concomitant failing of the Christian Democrats, his financial wealth and self-casting as an entrepreneur, his media empire, his ideological focus on "football imagery and anti-communism," and his charisma.

For some, Berlusconi's jokes have served an "undemocratic function" in that they foreclose dialogue by supplanting critique with humor (Edwards 2005, 238). Likening humor to the genre of myth, Michael Herzfeld has argued that Berlusconi's public offenses may "provide a means of suppressing the many inconvenient contradictions inherent in the social order" (2008, 146). Thus, Herzfeld suggests that one might read Berlusconi's jokes as staging yet simultaneously seeking to dissolve the contradictions surrounding the traditional versus the modern, rationality versus irrationality, or, more aptly, legitimate versus illegitimate governance. For historian Paul Ginsborg, the jokes furthered his "patrimonial authority," a concept developed from Max Weber's notion of "patrimonialism" and defined as a "personal, traditional authority [that] became more extended spatially and dependent upon different forms of interpersonal relationships [and] a reciprocity of favours" (2005, 118). Berlusconi, in this view, merges charismatic leadership with the regular manipulation of others' loyalty and indebtedness in fulfillment of his interests and the consolidation of power; put differently, his jokes serve as a strategic part of his "toxic" leadership, a poison to democratic institutions (Shin and Agnew 2008). His jokes seem to fit with a widely held view of him as a performer: "He is a showman who plays with . . . the collective imaginary" (Allum 2011, 290). Žižek (2009) concurs: "Behind [Berlusconi's] clownish mask there is a state power that functions with ruthless efficiency." These appellations—showman, clown—raise the question of what allowed such a figure of spectacle to become intertwined with the position of prime minister.

While concurring with the scholarly consensus that his humor has been significant, I focus here on excavating the cultural and historical conditions that made his techniques possible and effective: How was this curious intermingling of politician and masquerader a creation and function of his mediatized politics? Berlusconi's humor undergirds his oft-bewildering popularity because it forges ties to an Italian citizenry made cynical by the growing political spectacle of the late twentieth century: the carefully staged and sensational exhibitionism of national politics wrought in the 1980s and, in subsequent decades, the television media takeover of late-liberal politics. Berlusconi's self-crafted buffoonery, a form of manifest disguise, I suggest, fits well into larger and more deeply entrenched cultural narratives of failed but still charming funnymen and within an ethos of cynicism that valorizes possibly amoral but clever political leaders (Edwards 2005; O'Leary 2010; Watters 2011). I then turn to Antonio Ricci's news parody program *Striscia*, broadcast on Berlusconi's own Mediaset and one of Italy's most popular shows since it began in 1988 (Ardizzoni 2009; Chu 2012).[6] The most striking feature of the show is the character Gabibbo, who resembles a sports-team mascot: a human-sized, red puppet with a big mouth and rotund belly who is praised as a "civil defender." I illustrate how *Striscia*, especially the puppet Gabibbo, though apparently at odds with Berlusconi's own humor, is a twin effect: an opposing mode of political humor made possible by the same historical and cultural roots and tied to the technological mode of political communication: television. A few have argued that the show's programming has moved from a rather irreverent take on social injustices to a focus on gossip and celebrity and that, as a whole, it now remains relatively quiet about Berlusconi-related scandals (Ardizzoni 2009; Cosentino 2012, 58). But this is also what makes the show so vital in understanding Berlusconism: its literal fake news in a television system in which his economic control essentially rendered all of his news programming a form of disinformation.

Unraveling Political Humor and Cynicism

Oppositions and contradictions play a central role in humor, both for theorists (Bakhtin 1984; Douglas 1968; Freud 1990) and within an Italian tradition (Di Martino 2011; Pirandello 1974). On comedic performance, Gilles Deleuze has observed, "By scrupulously applying the law we are able to demonstrate its absurdity" (1971, 77). It follows, then, that political humor overdoes it: exaggerating political forms reveals the logics that undergird them (Boyer and Yurchak 2010). But humor is also about mockery and derision and, at times, embarrassing and ridiculing its targets. Embarrassing faux pas, Michael Billig (2001) suggests, do not necessarily produce empathy in onlookers, as humor theorists had initially conjectured. Rather, he argues, the enjoyment of such moments derives from their social mismatches, disruptions, or ambiguity (Italians laugh at the appalling distance between a refined political leader and Berlusconi; Billig 2001, 29). In turn, this pleasure is "disciplinary" and reinforces the social order by rendering those who laugh both happily transgressive and compliant (Billig 2005, 176).

The kind of humor disseminated by Berlusconi, I think, relies on a cynical national audience. It stems from a distrustful citizenry that finds amorality in politics to be dismally normal. The cynicism is reciprocal, as Berlusconi's own humor and political maneuverings exhibit contempt for traditional moral standards. Žižek (2009), for instance, has referred to "[Berlusconi's] rule through cynical demoralization." Catherine Fieschi and Paul Heywood have also suggested that cynicism—"a willingness to engage, but with lower expectations"—sustains Berlusconi's "entrepreneurial populism" (2004, 293). They define this form of populism as a media-heavy right-wing politics with business-based leaders and a less "xenophobic tenor" than "traditional populism," which is animated not by cynicism but "lack of trust" (Fieschi and Heywood 2004, 292). On the contrary, cynical citizens expect all politicians and leaders to be corrupt, conniving, and amoral, so, therefore, they come to value leaders' street smarts (*furbizia*) or cleverness and capacity to "play the game" (Fieschi and Heywood 2004, 303). In other words, Italians do have the ability to discern a political ruse, yet this backhandedness may be both widely expected and perceived as funny. The comparative record (Boyer 2013; Kendzior 2011; Wedeen 1999) reveals cases in which cynicism speaks to the much larger issue of hegemony: how constituents object to political authority but see it as inevitable and are, ultimately, compliant. For instance, according to Achille Mbembe and Janet Roitman, the crisis conditions of "do-it-yourself bureaucracy" (1995, 343), infrastructural failure, and uncertainty in Cameroon produce not only "confusion and chaos" (1995, 344) but also, from the public, "accommodation and acceptance of a *fait accompli*" (1995, 348) and possibly "protest by inertia" (1995, 350). The late-socialist subject, according to Alexei Yurchak (1997, 175), marshaled a "cynical reason" or "pretense misrecognition" whereby jokes about politicians, which circulated within unofficial spheres, showed that subjects recognized the falsity of official discourse even as they publicly complied with authority (see also Mbembe and Roitman 1995). Concurring with Žižek's (1989) earlier take on cynicism as a form of political ideology, Lisa Wedeen has examined satire and cynicism in Syria, calling it "the habituation to obedience—the combination of cynical lack of belief and compliant behavior" (1999, 154), which, she adds, may be found in both authoritarian regimes and Western liberal democracies. Berlusconi's own attempts at humor and his clownish behavior seem to be recognized in Italy as absurd, even funny, yet somehow inevitable. Take, for instance, journalist Francesco Merlo's (2011) prediction: "From now on [post-Berlusconi], jokes in politics will be remembered like a lifejacket for misfits, the ultimate safe house for the inadequate." Merlo cynically characterizes Berlusconi as a self-evidently deplorable leader—a misfit—and his humor as survivalist: the "lifejacket," keeping afloat what would otherwise sink.

I think it is also worth bearing in mind that the male figure has dominated the genre of Italian humor and comedy for several decades, or for centuries if we consider the tradition of sixteenth-century improvisational and theatrical comedy (*commedia dell'arte*; Rudlin 1994). Berlusconi's jokes implicitly reference a much longer legacy of masculinized and highly visible Italian clowns, with whom an Italian audience of onlookers can identify and yet ridicule. In the world of postwar cinema, the Italian style of comedy (*commedia all'italiana*) often centered on ridiculing "the ineptitude . . . of the

Italian male" (Bini 2011, 111). Since the early 1980s, this genre is also called *cinepanettone*, which refers in part to the Christmas cake panettone and the release of Christmas comedies but also to the mediocre men who star. Many of the celebrated works of comedian Alberto Sordi were satires of the emerging middle class and, in particular, a "childish, conformist, cowardly, irresponsible, and sly" man (Bini 2011, 136). While the films' comedic figures represented the flaws of Italian society, they also illustrated the viewer's own collusion in the country's social and political arrangements. During the 1970s, Italy's "leaden years" (*anni di piombo*) which were marked by left-wing terrorist violence and widespread corruption, these comedies "relied on the construction of a 'typical' but often grotesque Italian, usually male" who "allow[ed] the viewer no exit from his or her own complicity with the violence (of workplace, home or street)" (O'Leary 2010, 247). The protagonists typically pursue affairs and try to upkeep family and work, almost always subject to series of failures, mishaps, and rejections. In these genres, the exploits of these male protagonists rest on a figure of women as beautiful but without intelligence or agency. Like Berlusconi's programming on Mediaset and many of his own quips, women are systematically subordinated and fashioned as mere visual accessories (Benini 2012).

Some film critics have chastised Italian-style comedy for its ability to reduce citizens' desire to oppose or rise against particular political regimes.[7] If we understand contemporary political satire as riding on the coattails of this comedic tradition, it is hardly surprising that it shares the same double bind: the capacity to promote scrutiny of political life yet also dumb down political critique.[8]

Packaging Politics as Television Entertainment

Comparatively speaking, Italy's news parody programs emerged within a monopolized and highly politicized field of television broadcasting—such conditions bear resemblance to those underlying the emergence of other news parody programs such as *The Daily Show* and *The Colbert Report* (Boyer and Yurchak 2010), other forms of news parody across the globe (Baym and Jones 2012; Haugerud, Mahoney, and Ference 2012), and new forms of political irony (Haugerud 2012).[9] News parody is one genre within a late twentieth-century cultural shift toward greater media saturation and one in which televised media have played a strong role in shaping cultural identities, ways of being in the world, and new forms of desire and affect (Fiske 2011; Lukacs 2010; Mankekar 1999). *Striscia*, like other parody programs, emerged at a time when voicing political critique was particularly challenging, and its broadcast in 1988 represents an early emergence of televised political satire with respect to other Western democracies (Baym and Jones 2012, 5–6). So too Italian politics and politicians were thrust toward theatrical politics—where increasingly outrageous and preorchestrated public displays upstaged the content of political discourse—sooner and more dramatically than their counterparts in other Western democracies were (Smith and Voth 2002; Boyer and Yurchak 2010, 191). To understand Berlusconi's and *Striscia*'s heyday in the 1990s and 2000s, therefore, we must better parse two historical phenomena that converged in the 1980s: first, Prime Minister Bettino Craxi's corruption scandals and an emergent theatrical style of national politics and second, Italy's long-term and overt political manipulation of news programming and television.

"The Circus of Dwarfs and Dancers"

During the 1980s, the political landscape was marked by the dying embers of the Christian Democrats (DC), the unhinging of the Communist Party (PCI), and a massive corruption scandal that would culminate in the 1992 Tangentopoli (Bribesville) corruption trials (Edwards 2005, 227). In the center of this landscape was Prime Minister Bettino Craxi, a member of the center-left Italian Socialist Party (Partito Socialista Italiano, PSI) whom many Italians remember as the symbol of the corrupt decade. Importantly, however, in Craxi's leadership we begin to see slick media-ready stylings in the political arena, a valuing of clever political staging over political ideas (Smith and Voth 2002). Fellow party member and former minister of finance Rino Formica had famously called Craxi's entourage "a circus of dwarfs and dancers" (*un circo di nani e ballerina*) to mock the discernibly clientelistic ties between the prime minister and those who surrounded him, among whom were entertainers and showgirls (De Gregorio 1992). Moreover, those in his retinue dressed in high Milan fashion, and he once had a heart-shaped arrangement of carnations with the script "I love Bettino Craxi" landscaped into the grounds of his residence (Latella 2000). Marina Ripa di

Meana, one of his loyal supporters, said, "He recognized the value of jokes and thought it was inevitable that politics would drag a circus behind it" (Latella 2000, 7). To some extent, then, Craxi was a cunning crafter of showy politics; he knowingly fashioned political antics into a form of popular entertainment.

But Craxi could not fully marshal humor when the going got tough. When faced with corruption charges, his defense was to naturalize his circus-esque politics rather than continue the show. In 1992, during the Tangentopoli trials, he said, "Everyone is guilty, everyone knew" (Goffredo 1993). Craxi's sober admission represented a further betrayal of the Italian public, confirming, without any comedic allusion, that widespread corruption was germane to Italian political life. Favors and underhandedness, he said, were systemic, not the idle game of a single political performer. The Italian citizenry was becoming habituated, then, to simultaneously reckoning with the pleasures of consuming political shenanigans and disillusionment over irresponsible political figures.

Berlusconi's arrival in the mid-1990s, then, was greased by the "transformation of late-liberal politics into a kind of performance culture," one that prizes "staging, spinning, and images" and "formulaic political rhetoric" over political messages and ideas (Boyer and Yurchak 2010, 208–209). This too was indicative of a bigger Western shift in political humor that, by the 2000 U.S. presidential election, "illustrated a novel blending of politics and comedy seldom seen before in such magnitude" (Smith and Voth 2002, 124). In *Cracking Up: American Humor in a Time of Conflict*, Paul Lewis argues that "humor can seek to support ruinous policies and obscure personal defects" (2006, 194) and that being dubbed "humorless" has become a serious liability for a political candidate in the United States. Thus, Berlusconi did not just emerge within "the era of mediatization of politics" (Campus 2010, 227; Fieschi and Heyward 2004, 299), but as a media mogul and politician himself, he was, indeed, one of its innovators (Watters 2011, 171).[10] Recall that his 2001 comeback, with his glossy biography *Una storia italiana*, produced an "electoral marketing blitz [that] played on his own celebrity" (Edwards 2005, 227; Watters 2011, 173).

Craxi, however, paved the way for Berlusconi in several ways. Craxi presided over the "start-up" version of "mediatized" politics: he left the Italian population both primed for political spectacle and more cynical than ever. He also helped Berlusconi judicially by clearing the path for commercial enterprise to enter national broadcasting and, as a result, for Berlusconi's Mediaset network, financed by his company Fininvest, to become a media power in Italy (Edwards 2005, 236); the law Craxi engineered would later be known as the "Berlusconi decree" (Allum 2011; Hibberd 2007, 886). One could interpret Berlusconi's 2003 inquiry into the Tangentopoli judiciary as a way of returning the favor to Craxi, to whom he owed much.[11]

News Parody and Ricci's Satirical Genius

Since the inception of Italian televised news programming in the 1950s, each channel had been devoted to one party's message, part of a much-larger system of *lottizzazione* (Briziarelli 2011, 13), or "sharing-out" of political power (Edwards 2005, 227). In the period from 1948 to 1993, the Radiotelevisione Italiana (RAI) channels were distributed along the following party lines: RAI-1 for the Christian Democrats, RAI-2 for the Socialists, and RAI-3 for the Communist Party (Domènech 1990, 72). The veracity of news reporting was thus dubious and questionable in Italy for several decades before *Striscia*. By the late 1980s, broadcasting had become "duo-polous": split between the predominately public-service RAI and tightly owned private networks, headed by Berlusconi (Cosentino 2012, 53).[12] It was this bifurcated system that made the show's initial attack on the legitimacy of public broadcasting news an especially effective move (Cosentino 2012, 55).

Informed by Antonio Gramsci and Marxist theorists, show creator Antonio Ricci (1998) has aimed at combating the "ideological" dimension of Italy's public television (Cosentino 2012, 55–56). Ricci has been no stranger to controversy. He has called television "a window on the market not the world" (1998, 8) and a "monster maker" (1998, 58) and insists that news media "sell you a political idea" (1998, 8). Berlusconi has even challenged the creator directly: "Good always triumphs over evil, except in the case of Antonio Ricci" (Ricci 1998, 39). His work fundamentally altered Italian television, which, at least prior to Ricci, was best known for its vapid variety shows, its continual staging of the sexualized female body, and the marginal space it offered to ethnic and racial diversity (Ardizzoni 2005). Ricci had already found success with several programs, such as *Drive In*, which aired from 1983 to 1988 on Italia-1. In *Drive In*, early forms of news parody were already present: Ezio Greggio, for instance, who later hosted *Striscia*, performed a skit called

"Spetteguless," a play on the word for gossip, with the slogan "Not news chronicles, express stories . . . more than news, gossip."

In another of his shows, *Fantastico* (Fantastic), which aired from 1979 to 1982, Ricci began working with comedian Beppe Grillo, who has since become a political force in Italy. On one 1986 episode, Grillo joked about how then prime minister Bettino Craxi and Socialist Party members stole from one another. In the show *Odiens* (Audience), Ricci featured various strategies for civil disobedience, such as avoiding telephone fees and traffic tickets and reusing postage stamps (Tanzarella 1988).[13] It was no coincidence that Ricci's politically shrewd artistry was brewing during the 1980s, a time when Italian politics was becoming more corrupt and media dependent.

With Berlusconi's rise, however, broadcasting faced new challenges to fund and produce shows on non-Berlusconi-owned channels or, at times, to get airtime for shows that criticized him directly. Some analysts suggest political satire in Italy actually increased in production and popularity (Ardizzoni 2009; Ferrari and Ardizzoni 2010). Television also became a site of overt censorship, with Berlusconi directly attempting to silence political satirists whom he accused of "making criminal use of public television" (Edwards 2005, 235; Hibberd 2007; Watters 2011).[14] In 2002, for example, satirist Daniele Luttazzi was kicked off Italian airwaves and subsequently made a DVD subtitled *Bin Laden Can Get on TV, but I Can't* (*Economist* 2004). Berlusconi impersonator Sabina Guzzanti could only get her 2003 program *RaiOt* aired in the dead hours of the night, and Mediaset, Berlusconi's TV conglomerate, attempted to have her indicted for libel (*Economist* 2004).[15] The same year, Italian filmmaker and Nobel Prize winner Dario Fo produced a satire about Berlusconi, *L'Anomalo bicefalo* (The two-headed anomaly), and he too faced challenges finding airtime for it. The satellite television network Sky Italia eventually showed it but without sound or subtitles (Watters 2011). For political satire to survive, it had to be creative and savvy and, more importantly, explore alternative forms for its message.

Puppet as Truth-Teller

We can now, finally, better understand the news parody of *Striscia* because of its precedents: the turn toward politics styled and tailored for television media, a troubled yet ardent production of political satire, and a cynical

Figure 1.2. News parody program *Striscia*'s Golden Tapir Award (Tapiro d'Oro). Photo by Roberto Di Cristina.

citizenry. *Striscia* airs at dinnertime Monday through Friday, and numerous Italians have come to deeply associate the setting of the table and the evening meal with *Striscia*, which is not coincidental: Ricci calls it the "telemass . . . for Italy who eats in front of the TV" (Greco 2011). The show features two anchormen and intermittent dance acts by two now-iconic female performers known as the *veline*; a variety of (real and fake) news segments with correspondents; footage of real newscasts and programs; and outtake footage of newscasts (*fuorionda*, or "off air"; Cosentino 2012).[16] A 2010 outtake reveals a scene of political deal making in which the mayor of Latina, Vincenzo Zacchero, reminds the Lazio region governor, Renata Polverini, to "remember his daughters" in return for his having secured her over fifty votes (Cosentino 2012, 62). *Striscia*'s outtakes bring this type of political corruption directly, and regularly, before the eyes of Italian viewers.

Striscia frequently reminds the public of Berlusconi's idiocy with "straight" news reports on his gaffes: how, on three different occasions, he rearranged world leaders from Montenegro, Serbia, and Russia for an official photograph because the men were all significantly taller than he (*Striscia la Notizia* 2010a); how he took a nap during a press conference (*Striscia la Notizia* 2010b). Some of the cleverest satirical bits involve the show's mock award, the Tapiro d'Oro (Golden Tapir). Many were awarded in relation to Berlusconi's trial and the ongoing scandal in 2011. For instance, RAI anchorman Emilio Fede received a special "Bunga Bunga Tapir" after he reportedly attended one of Berlusconi's sex parties and was accused of aiding and abetting prostitution (Hooper 2011; *Striscia la Notizia* 2011d).[17] In January 2011, another skit included a song to the tune of Laura Branigan's "Gloria," with Fede repeating "bunga bunga" as Berlusconi sings, "I'm sorry but I'll continue to do what I want in my house" (*Striscia la Notizia* 2011b). A "giant tapir" (*tapiro gigante*) went to Berlusconi after he resigned in November 2011. Refused access to Berlusconi's political headquarters in Rome to deliver the prize, the *Striscia* news anchor was momentarily stumped, and then had a "eureka" moment: he applied red lipstick and a blonde wig to the tapir statuette, proclaiming, "Here's our 'escort' tapir! We transformed it into a beautiful tapir-esse [*tapiressa*]. . . . Maybe with her [dressed up] like this, we'll be let in!" (*Striscia la Notizia* 2011c).

Afterward, one of the program's two hosts, Enzo Iacchetti, joked about how the tapir was nearly smuggled inside the political barriers: "It's kind of a Trojan tapir [*tapir di troia*]." Here the joke was through the pun on "Trojan," which can refer to both the Trojan horse and a prostitute. The other host responded, "Well, in situations like this, people'll just take the first girl they get," and then, after a well-timed pause, he added, "These *are* crisis times" (*Striscia la Notizia* 2011c). In these sketches, Berlusconi becomes the object of collective ridicule; rather than identification with the ineptitude of the leader, they prompt sentiment against him. One might say it is easy to make fun of politicians' sex scandals. Yet in these savvy one-liners, the threads of affinity that Berlusconi himself works to forge with the Italian public are perhaps not severed but frayed. Alternatively, Berlusconi sympathizers might find the parody funny but still identify with what the skit also reaffirms: his oversexed masculinity and levity. At the same time, the image of the oversized bronze tapir, ornate with sloppily applied red lipstick and a blonde wig, is vulgar: popular and lewd. "The production of vulgarity," Achille Mbembe tells us, "needs to be understood as a deliberately cynical operation" (1992, 16); that is, the "lies and double-speak" that he suggests characterize postcolonial authority and communist regimes, and their effects bear a resemblance to millennial regimes of power in Italy.

In 2012, *Striscia* was the most-watched television program in Italy, with a 20 percent share of the prime-time audience (AGI 2012); it was awarded a place in the *Guinness Book of World Records* for the longest-running satirical news shows (Guinness World Records 2012). I asked some *Striscia* viewers to explore why they enjoyed the show. Francesca, a divorced working mother in her fifties who watches *Striscia* regularly, told me she enjoyed the show because it was "easy-going but also interesting" (personal communication, September 20, 2012). She continued, "I especially like the parts tied to often-overlooked Italian happenings (buildings in disrepair, scams, etc.) and most of all I like the fact that the outtakes come back to the same cases to keep viewers up-to-date on the latest developments." While she suggests the show has not transformed her political views, she admits it has made her realize "the political situation was more devastating than [she] had believed." Roberto, a psychologist and poet in his late thirties who produces his own satirical YouTube videos, by contrast, calls *Striscia* "a 'distraction' because it never hits the true issues and evades the solution which are all 'delegated' to the latest puppet to hold office" (personal communication, September 7, 2012), cynically hinting at the latest elected leader. For Roberto, *Striscia* fans represent a segment of the "public with a low cultural profile [who are] the average Berlusconi voters." I asked him to elaborate on what he meant by a "low cultural profile." He explained,

> The average voter is uninformed, holding low educational degrees, very old or very young, and uninformed about politics. The key point, I think, is that these people get their information from TV (which, in Italy, is hardly pluralist) and, above all, from the three eminent Berlusconi TV stations. . . . Thus the average voter is not representative (as they would like us to believe) of the productive bourgeoisie of North Italy, nor of entrepreneurs, nor of great economists and intellectuals. . . . *Striscia* is a show that "contributes" to this by bringing attention and favors to Berlusconi because he can be perceived as interesting, informed, productive, and fun. . . . In order to appreciate *Striscia*, you have to believe that the Berlusconian government is capable of resolving problems: that you must give them trust, because they know what to do. . . . At the most, you can report a problem to [the government] and watch if they resolve it, but everything is a show: produced and filmed just like a show. They don't know how to resolve problems, but they'll *show* you that they have. . . . In general, *Striscia* shows problems that relate to particular situations, small things, and at times, personal issues. . . . Often *Striscia* unmasks

> fake healers and tells viewers not to trust them . . . (at least this is an effective action from the legal standpoint but remains within the scope of persecuting the individual not the system).

Roberto's savvy critique of *Striscia* shows us that Italian citizens can read against the grain and see the show as indirectly supporting Berlusconi not only in its individualization of social problems but also in its capacity to render him "interesting" and "fun." After all, the efficacy of what we might call "showman populism" depends precisely on this image; and, even if *Striscia* intends to reveal his tactics and leadership as asinine, it nevertheless re-creates Berlusconi as "fun" in ways that, at least sometimes, sustain and bolster his leadership and authority.

In Roberto's comments, we also find his indictment of, and weariness with, politics as "a show" and political leaders as "puppets." Puppets, in turn, return us to the most memorable and provocative character in *Striscia*: the

Figure 1.3. News parody program *Striscia*'s puppet Gabibbo in Turin in 2008 at an event for Italian Union of Parents against Children's Tumors (Unione dei genitori contro i tumori dei bambini, or UGI). Photo by Antonio Scardinale.

mascot Gabibbo, introduced in 1990, a tall, red puppet with a big, bulbous head and body, whom creator Ricci calls "the most credible journalist in Italy" (1998, 18) and who is widely seen as such (Cosentino 2012, 53; Panarari 2010, 9). Yet viewers not only trust the puppet as an honest reporter but also turn to him as a political mediator: the "SOS Gabibbo" hotline exposes fraud, local political issues, and corruption (Cepernich 2008; Cosentino 2012, 56). Fans can submit e-mail messages reporting "injustices, scandals, or scoops" on Gabibbo's website. In February 2011, for instance, Gabibbo, who regularly defends un- and underemployed workers, came to "intervene" when four hundred of eight hundred factory workers were at risk of losing their jobs because of the company's plan to begin offshore production (*Striscia la Notizia* 2011a). Nicknamed the "Red Avenger" (Vendicatore Rosso), Gabibbo often makes such visits. In 2009, he visited an abandoned prison and criticized prison overpopulation, offered solidarity to the residents of a flood-devastated town in 2010 and the home of poor immigrants in 2007, exposed fraud at the 2006 Torino Olympics, and, in 2004, lamented the rising price of milk (*Striscia la Notizia* 2012a). For Ricci (1998, 59), Gabibbo is as "monstrous" as television itself because he makes transparent a social truth: only those in masks can effectively reach audiences, or, put more darkly, "the puppet becomes real" (1998, 134).

Striscia, with its mascot Gabibbo, has had a significant effect on Italian political culture. It is considered a legitimate and reliable site for news and political watchdogging by a broad spectrum of institutions and groups. In November 2011, the show was honored by Milan's prestigious Bocconi University for its work; the university called Gabibbo a "civil defender" and credited the show with divulging over fifty-eight billion euros (and as much as 140 billion) of "wasted money" spent on corruption, scandal, or crime (Salvaggiulo 2011). This is but one of numerous awards it has won. Others include a 2006 award from the Italian Police Association for being the "most important show on social issues," a 2007 award from the National Association of Occupational Mutilated and Invalid People, and a 2011 award from the Financial Guard (Guardia di Finanza) for its commitment to legal action (*Striscia la Notizia* 2012b). *Striscia* is also commended for prompting 174 investigations on behalf of consumers, tens of arrests, and civil cases (90 percent of which plaintiffs have won; *Corriere della Sera* 2012; *Striscia la Notizia* 2012b).

At first glance, Berlusconi and Gabibbo seem entirely unrelated. Berlusconi, of course, wears no giant red suit like Gabibbo, but like the madcap red puppet, he does not conceal his spectacle; his phoniness is apparent. An ethos of cynicism, together with the rise of theatrical and mediatized politics, has made unhidden artifice—in which spectacle is fully undisguised and overt—appealing or humorous in Italy. And so we find a similar cynical logic shared by Gabibbo fans and Berlusconi supporters: they prefer unconcealed masqueraders because they assume that people in power (politicians or news reporters) who do not outwardly display artifice are covertly wearing invisible masks.[18] Roberto, neither a *Striscia* nor a Berlusconi fan, it follows, finds overt public fandangle, whether in government or in the media, devoid of integrity and potential for systemic change and, thus, repugnant. One might say that figures in disguise have been in a privileged position of truth-telling for centuries in Western imaginaries (the court jester and those who populate the masquerade ball, Shakespeare's plays within plays, and the carnivalesque are examples). There is a precedent too of distrusting over-the-top political spectacle, as in "Pay no attention to that man behind the curtain." Who says Gabibbo's red bulk cannot hide a puppeteering patriarch? Still, the popularity of Gabibbo and Berlusconi, though the former is trusted and the latter disdained, shows us the range of figures that can surface from an ethos of televised theatrical politics spiked with an undercurrent of cynicism. Though Gabibbo and Berlusconi command popularity in different ways, they both owe their success in contemporary Italy to the appeal of visible artifice. Though cynicism may produce the citizenry's general mistrust, it also, strangely, yields a preference for manifest, as opposed to hidden, disguise.

Comedy and Confusion

In *Mamma Mia: Berlusconi Explained for Posterity and Friends Abroad*, essayist Beppe Severgnini quotes Berlusconi as saying during his 2011 prostitution and sex-party scandal, "The Left keeps telling people 'send Berlusconi home,' which causes me a certain embarrassment. Because I have twenty homes, and I wouldn't know which one to go to" (2011, 16). The most powerful man in Italy makes his removal from political office into a joke: "I can't help being rich and naughty." Gilmour (2011) and Severgnini (2011) agree that part of Berlusconi's success has been his ability to embody and behave like the "average" Italian. Severgnini puts it plainly: "He prevails by complicity. . . . I am like you: impulsive, intolerant, enthusiastic, and impatient" (2011, 13). His humor invites Italians to identify with him—his wealth, his business savvy, his ambition, and irreverence—and, in doing so, to collude with his politics. It aspires to a governance of shared laughter and forgiveness, in which Italy's cynical citizenry assumes his amorality and corruption yet finds amusement in how he weasels out of any responsibility for his behavior. Yet this amusement is hardly apolitical; laughter is intertwined with anxiety, uncertainty, and fear, conditioned by an abiding acrimony and disenchantment. As Mbembe and Roitman wisely remind us, "Fear, and the laughter it provokes, are often an effect of the ambiguity of lived experience: one is subject to violence and yet, often in spite of oneself, one participates in its very production" (1995, 353).

The past few decades have shown far-reaching innovations in how humor and irony emerge from politics and how politics deploy humor (Baym and Jones 2012; Dmitriev 2006; Fernandez and Huber 2001). Interestingly, however, some of this comedy is deeply intertwined with the participant's confusion, the viewer's flummoxing. Angelique Haugerud (2012, 146, 154) has examined the U.S.-based protest group "the Billionaires," who, while dressed as the wealthiest elite and holding signs such as "Widen the Income Gap," stage a kind of political satire that can sometimes befuddle spectators. Haugerud suggests the Billionaires "denounce inequality through empathic humor" (2012, 154); the group builds from a love–hate connection between the billionaire and the everyday citizen, in that elitism and wealth become the objects of derision but also identification. In examining the U.S. news parody of Jon Stewart and Stephen Colbert, Boyer and Yurchak (2010, 181) chart a similar confusion in *stiob*, a form of humor characterized by "overidentification" that results in ambivalence because of its hypernormative adoption of the conventions of political discourse. On *The Colbert Report*, for example, Stephen Colbert "performatively almost never steps out of character," which in turn makes media pundits "often uncertain how to engage him" (Boyer and Yurchak 2010, 195). Consider Berlusconi's joke about how he would be "sent home." These comparative examples help us scrutinize how Western political culture destabilizes citizens' ability to distinguish the premeditated from the spontaneous, the artificial from the genuine, and the scripted from the heartfelt. Subjects thus continually reproduce the conditions that enable this comedic confusion and, in the most cynical of foretellings, may have further trouble distinguishing their own affective states and interiorities. Just as the demand for emotional labor that emerged in postindustrial economies produced emotional confusion in the laborer (Hochschild 1983), so too theatrical political culture, together with hyperbolic valuing of autonomy, may affect how subjects understand the world around them and their psychological and affective registers.

Chapter 2

The Soldiers of Rationality

For over twenty years CICAP has been committed to combating irrationality, superstition, and prejudice with the weapons of science and reason.

—CICAP

On June 24, 2011, the evening of the summer solstice, citizens of the northeastern Italian city of Vicenza were invited to the piazza to celebrate "Witchnight" (La Notte della Streghe). Witchnight came on the occasion of the feast day of Saint John the Baptist. The city's promotional flyer showed the dark silhouettes of three witches dancing in the moonlight under a tree. The evening promised a spectacular joining of earth's elements: "The sun (fire) marries the moon (water) and the bonfires (Saint John's fires) will be lit in the fields, which is appeasing and purifying." Grass collected on that night would have a "special power" to "get rid of any illness," and "keep the bad spirits away." Unmarried women eager to predict their marital futures were invited to break an egg into a glass of water and leave it next to the window overnight to absorb Saint John's essence. Finally, farmers were prompted to tie a special braided cord around their walnut trees so they might produce sweet walnuts and excellent walnut liquor (*nocino*). Across Europe and for many centuries, Saint John's feast day has blended pagan celebrations of the solstice with the Catholic tradition of saints. The event has often highlighted visits from witches and demons alongside agricultural rituals promising good health and wellness.

Meanwhile, a week prior, the Committee for the Investigation of Pseudoscientific Claims (Comitato Italiano per il Controllo delle Affermazioni sulle Pseudoscienze) staged a rally in June 2011, "A Day against Superstition" (Una Giornata Anti-superstizione), held in various Italian cities, including Rome, Vicenza, and Pisa. The group's national secretary, Massimo Polidoro, outlined the organization's goals: "For over twenty years CICAP has been committed to combating irrationality, superstition, and prejudice with the weapons of science and reason" (CICAP 2011).[1]

What do Italians talk about when they talk about scientific reasoning as the arbiter of truth? What is at stake in the nonscientific for scientific skeptics, especially in this fraught political and economic context? Though beliefs in supernatural and occult orders have long been intertwined with early Catholic practices, supernatural ways of knowing have become more overt and pronounced in the latter quarter of the twentieth century: traditional Catholic Church affiliations have steadily declined since the 1970s and Italians have increasingly moved toward belief in the supernatural, ranging from witches to aliens to Catholic mysticism and exorcism (Trocchi 1990, 1998). Scientific activism—science-based social movements from rationalist movements to atheistic and skeptic movements—have also been increasingly widespread since the late 1980s in Italy and around the world (Novella

Figure 2.1. CICAP's "A Day against Superstition," June 2011. Photo by author.

and Bloomberg 1999; Nisbet 2000; Quack 2012). The question becomes what aspects of Italian life give rise to these seemingly opposing modes of thought.

The Comaroffs (2000) have famously theorized millennial capitalism: neoliberalism as giving rise to forms of, after Michael Hardt, "casino capitalism"—gambling, Ponzi schemes, get-rich-quick plans, and in short, "magic money" (298, 292). Risk-bearing economic life engenders occult beliefs, they theorize, because of the way economic booms and busts appear simultaneously invisible, unpredictable, and magical. The strange appearances of money-for-nothing gives rise to new religious movements as well as supernatural panics from zombies to witchcraft to body snatching rumors around the globe (Comaroff and Comaroff 2000, 314). But our contemporary material infrastructure also produces the apparent opposite of magic and its fantastic cronies: "hyperrationalization," which has been, I would add, undertheorized and underdocumented over the past decade or so (292). In this sense, my interest in "the enchantment turn" in Italy may run parallel to the neoliberalizing of Italy's economy, its siphoned secure labor, and citizens' uneasy reliance on state and social services (318). After the past twenty years of highly rapid economic reorganization, we find mass unemployment, or a precariat class, especially among youth, insecure leadership riddled with corruption and scandal, and national politics at its most spectacular and media-saturated. Together with its underlying Catholic epistemology, Italy's socioeconomic regime might represent a catalyst for both tighter and looser empiricism.

Uncertainties in the foundation of empirical knowledge, including sensory knowledge, witnessing, and testimony that are both highly specific to Italy, yet also characteristic of late modernity, might compel modern subjects to both embrace the supernatural and hold fast to rationalism, or the kind of radical empiricism now virulent among Italy's skeptics. But before we move to the contemporary context, we must first examine how Italian superstition, together with the emergence of scientific thinking, rumbles along the fault lines of paganism, Christianity, and Enlightenment thinking. In order to understand "anti-superstition," therefore, means first asking what superstition has been threatening, that is, which grounds—epistemological, experiential, sacred, and political–it has shaken.

Global Movement of Skeptics and Rationalists

In addition to hosting the 2004 World Skeptics Congress, Italy's Padua-based members of CICAP might be considered an engaged part of the global skeptics movement.[2] CICAP has over ten centers in Italy. It is affiliated with the European Council of Skeptical Organizations, the Committee for Skeptical Inquiry in the United States, and global events such as the Sixth World Skeptics Congress in May 2012, "Promoting Science in an Age of Uncertainty." The conference was organized by "The Skeptics," self-defined as "an international network of scientists and scientifically-minded individuals whose aim is to examine claims and theories that lie at the fringe or even outside of our best current science, and to inform the public about the results of these investigations" (World Skeptics 2012). Taking on creationism, alternative medicine, and the occult, this global network of skeptics promotes a worldview mediated by their belief in objectivity, rationality, and empiricism.

Italy's CICAP is part of a larger global network of science-based movements organized under various labels from rationalism (Quack 2012) to scientific skepticism (Novella and Bloomberg 1999). In India, Johannes Quack (2012) has investigated the Organization for the Eradication of Superstition (ANIS) and "the wider international movement that advocates a secular, materialistic, and naturalistic worldview" (4). Two American members of skeptical societies, Steven Novella and David Bloomberg (1999), members of the New England Skeptical Society and the Rational Examination Association of Lincoln Land, respectively, distinguish "scientific skepticism" from rationalism: whereas skeptics focus on "testable beliefs," not religious beliefs, rationalists adopt a materialist scientific approach toward anything that "lacks objective evidence," including religious phenomena (45). While skeptics focus on phenomena without a long tradition of scientific examination such as paranormal abilities like mind-reading or what they see as outside standard religious views—such as a bleeding statue of the Virgin Mary—rationalists consider the omission of religious beliefs from the purview of scientific study antithetical to their worldview and "intellectually dishonest" (45).

Catholicism, often seen as the gateway to engagement with the supernatural world, was an underlying but explicit feature of the Day against Superstition (Csordas 2007). Skeptics regularly distinguish between "reasonable" Catholic piety and what they see as its exploitation: they made pamphlets on the Shroud of Turin, bleeding statues of the Virgin Mary, and local miracles for passersby. Novella and Bloomberg (1999) aim to make this distinction between religion and negotiable religious practices for skeptics: the claim that God created a world that included evolution over billions of years must be, they insist, "banished to a realm outside of science" (45). Rather than distinguish between religion and science, they suggest that the core divergence exists in the difference between faith, that is, beliefs not based on evidence, and science, that is, evidence-based beliefs. As skeptics see it, one may have this type of faith in various kinds of paranormal and New Age actors and phenomena such as witches or aliens just as one can have faith in religious deities. Such a distinction allows them to produce a new ontological category, "testable religious claims," such as "creationists, faith healers, and miracle men" and "weeping icons" which come under the purview of scientific skeptics (Novella and Bloomberg 2010, 45–46).[3] Novella and Bloomberg's ideological distinction helped me understand why Italy's CICAP also investigates phenomena such as faith healers but not other tenets of Catholicism: they counted as "testable" religious claims for skeptics. When I asked, for instance, whether the pamphlet available on the Day against Superstition on the Shroud of Turin meant CICAP aimed to debunk Catholicism, I was met with slightly awkward silence. "We investigate only the overt extrapolations of Catholic faith," I was told.

The dichotomy between what skeptics call faith, poor-evidence belief, and religion, impossible-evidence belief, presupposes that evidence is binary: either existent or nonexistent; ignoring how evidence exists on multiple spectrums of plausible and implausible, visible to nonvisible; repeatable versus nonrepeatable. Like evidence, the nature of scientific knowledge is also partial (Latour 2004); as Sabina Magliocco (2012) asserts, "Belief does not require the acceptance of a coherent set of propositions" (10) and acts as "a state of conviction" coexistent with rationality (12; see also Tambiah 1990). Scientific discourse on belief and skepticism, however, regularly reinstates a more simplistic bifurcation of evidence. Richard Brewer (1992), a scientist writing for *BioSciences*, published an essay on "how people construct erroneous ideas" (123). Credulousness, he argues, deploying some armchair evolutionary theory, must "increase fitness" (123). However, he concludes that "layman's skepticism" toward all expert knowledge, whether magic or science, would be the most "adaptive" kind of behavior: nothing would be blindly accepted without proper scrutiny and judgment. The philosophical notion of skepticism animates Brewer's thinking: a purer sense of doubt, a desire to scrutinize claims about the world. It is precisely this definition of skepticism that scientific skepticism, if examined more closely, actually subverts. Indeed, this form of skepticism dwells on a spectrum of doubt that may be part of multiple—empirical, immaterial, religious, occult—epistemologies yet is haunted by another false divide: rational doubt and irrational doubt (Shanafelt 2004, 328). In this case, the former suggests normative engagement with evidence, while the latter implies a pathological, even paranoid, procedure for evaluating claims.

Science and "Junk Science"

In June 2011, I was hanging out at the CICAP's center on the outskirts of Padua's historic center, browsing CICAP's library, and chatting with members Mario and Lara about their interest in the group. Mario recounted how they had participated in a national event, "Day for the Correct Scientific Information" (Giornata per la corretta informazione scientifica). Mario had attended a few panels, one on the need for animal testing in medicine, which was organized in response to a radical animal rights group that had freed some hamsters in a Milan lab. Another talk was on the safety of genetically modified foods, which tried to assuage fears about the "Frankenfoods." Ignorance is at the root of these problems, Mario explained, especially when grain was a hybrid of three different plants. Later, we went to a talk by Salvo di Grazia titled "Does Alternative Medicine Exist? How to Distinguish Science from Junk Science" to a mostly university student audience of about fifty. Di Grazia's talk sought to debunk homeopathic medicine and persuade the audience that homeopathy was junk science, which he said was not evidence-based medicine but rather supported by word of mouth, anecdotal testimonials, and false advertising. For false advertising, he gave examples of standing ovations and false credentials for practitioners and spoke about how incredible stories of healing preyed on people's vulnerability and suffering. To demonstrate the false credentials of the practitioners, he showed examples of his own fabricated requests to go to a Chinese homeopathic conference. He called himself "Massimo di Serieta'," which is a joke name of Maximum Seriousness, and a paper on how to make a solution based on the ingredients of pasta carbonara—egg, cheese, pepper, and guanciale (an Italian cured pork product). Not only was his paper accepted at the conference, he said grinning widely, but he was also named the chair of the panel. He got a huge laugh.

At one point his slide asked the audience pointedly: "Sei paranoide?" (Are you paranoid?) and most people laughed and nodded. Here, he was showing that fake medicine and fake science used questions designed for an affirmation as a technique. He also said we must be watchful for elaborate demonstrations that would try to convince us of the efficacy of a medicine or procedure but were actually fabricated. He took out a large bottle with a vodka label, which he said he made using the homeopathic technique, diluting one drop in water many times over. To demonstrate further, he asked his eight-year-old son, who was in the audience, to come up and drink a glass in front of everyone. "It's just water," he told us, winking at the audience. The son ran merrily down from the top seat, drank the glass in one big gulp, and then smiled at his dad. The audience clapped as he ran back to his seat.

At the end, he asked for questions from the audience. One woman said she had an autistic child and asked whether she should try a cure, if only because her autistic daughter might respond because of the placebo effect. She framed her question as hypothetical, even as she seemed to be speaking from experience. He said that while her daughter might have a placebo response, the mother's knowledge of the placebo would also influence her daughter's well-being and her perception of her daughter's illness. Here, he seemed to reference double-blind placebo trials where the doctor's knowledge of whether the doctor was giving a placebo or actual medication to the patient also influenced the patient's healing. Another person asked why judges and the legal system were not getting more involved in criminalizing homeopathy. "Well," he said, "selling water was not a crime," adding, "judges are not scientists." Another woman told a brief story about how she had used an experimental algae treatment to good effect on her after ten treatments, even though "it didn't really convince me." The questioner had properly shown her, as she was effectively socialized into this climate, skepticism toward the nonbiomedical treatment. He was polite but still quite firmly told her that our own understanding of our beliefs—even our skepticism—would not change the placebo effect. He referenced a study that found that prayer was more effective for healing atheists than for believers in God. "We have strange minds," he said. Taking a more somber tone, he warned the audience: "We're living in strange times and we must defend science. Pulling back is admitting defeat."

Just as we audience members were lured to questioning our own responses to alternative medicine, so, too, we were invited to see such practices as part of a growing and coordinated war against science. The questioners also modeled a kind of bodily discipline, in which nonscientific treatments might be framed as placebo responses, and the possible response was about fine-tuning our understandings of how to best explain results. Of course, the discourse of placebo and nocebo has long been a scientific method for rationalizing and scrutinizing bodily responses to a host of nonbiomedical remedies, not to mention witchcraft and shamanism. Here, too, we were offered a pedagogy on how to use rationality and logic to gauge our body's response.

The Superstitious Other

In 2011, CICAP's leader Massimo Polidoro (2011) commented on the goals of the demonstration against superstition: "Being superstitious is itself unlucky. Believing that an object, a person, or a phrase has the power to generate disasters is a self-fulfilling prophecy." Polidoro translates superstition into a psychological deficiency, a cognitive sublimation of risk, a reading practice for reality capable of sustaining any possible outcome. A 2010 newspaper reported that the number of Italians who admit to believing in superstitions is at nearly 60 percent and "at the height" compared with the European average (*City* 2010). This is one instantiation of familiar orientalizing discourse in which Italy's supernatural beliefs became Other to European modern subjects. In 1864, historian Camillo Mapei wrote, "But is it not true perhaps that superstition reigns in Italy? . . . As imagination is more lively and indolence greater, superstition exerts a greater influence over the people" (Mapei 1864). Another mid-nineteenth-century historian noted that in Italy he found "the lower classes superstitious and the higher classes lovers of pleasure" (Ramage 1868, 1). He asked, seemingly bewildered, "What can be done with a people in this abject state of superstition? What effect would a more spiritual form have upon them?" (Ramage 1868, 112). Italy, particularly the South, has long been characterized as superstitious by observers for centuries in ways that brought into being the credulous Italian subject: a childish, unreasonable, sometimes paranoid believer whose ritual practice habitually engages with and finds meaning in visible and invisible objects, beings, and worlds (Moe 2002).

The overwrought nineteenth-century progress narratives of savagery and civilization were easily mapped onto supernatural beliefs as opposed to religious, though largely Judeo-Christian, orders embedded within European imperialism. During this period, the idea of superstition implied abjection, laziness, and possibly heresy. Renowned expert on magic in Italy Ernesto de Martino (2004 [1959]) argues that it was a uniquely Protestant and anti-Church discourse that pitched Italy as endlessly "superstitious" because the Church's hold was seen as merely a façade to an otherwise pagan population (124). Yet the basic idea that Italian religious and spiritual beliefs were shaped by early Christianity is not a false one, or only rhetorical, and is a deeply enduring discourse. In his collection of letters written between 1846 and 1850, for example, French philosopher Ernest Renan visited Naples and wrote, "Religion is nothing but pure superstition, the expression of fear or self-interest" (Moe 2002, 74). In 1763, Voltaire figured superstition as the "mad daughter" of religion.

The proximity between the supernatural and Catholicism was established quite long before the Protestant Reformation. According to historian of European superstition Michael Bailey (2007), the rise of Christianity in the fourth century combined beliefs and practices in magic, witchcraft, and superstition that were already present in early modern Europe. In 789 Holy Roman Emperor Charlemagne outlawed all magical practices with the General Admonition (Admonition Generalis), and practitioners were killed as punishment. Magic, into which superstitious practices were often folded, was defined as both separate from and counter to Christianity and, specifically, as a methodology of employing supernatural and occult power to change the world in ways other than by direct physical means (Bailey 2007, 11). By and large, this produced a lasting ideology. It divided the world into two domains: the demonic and the divine (Bailey 2007, 12). However, these two realms were understood as ambiguously intertwined: to the ancient Europeans both demons and gods might be called on for help or harm, even as practices developed to condemn what we would now call magic or witchcraft. Only in the early Christian period did forms of magic become "condemned as inherently superstitious, pagan, and incommensurate with the Christian faith" (Bailey 2007, 51). In the next several centuries, Christian authorities were even imagined as supporting beliefs in magic, or might struggle to determine whether something was influenced by demonic or divine authorities.

The rise of humanism, the Protestant Reformation, and rise of intellectual movements during the renaissance periods of the fourteenth and fifteenth centuries did not eliminate the belief in and the practice of magic and witchcraft on the continent (Bailey 2007, 180). Italy had the earliest number of witch trials of mostly young women in the fifteenth century and over two thousand executions in the sixteenth and seventeenth centuries. In fact, in the sixteenth century, certain spells and acts of conjuring were called superstitious by the Church if they did not involve God, yet other similar practices, such as "curing rituals" by authorized persons, were acceptable (Sluhovsky 2007, 34). This multicentury ambiguity between piety and superstition was not resolved, which led leaders of the Italian Enlightenment in the seventeenth century to accuse the Church itself of "perpetuating superstition" (Killinger 1992, 86).

In the eighteenth and nineteenth centuries, rural Catholicism was mixed with various practices of superstition, magic, and paganism (Duggan 1984, 26). Furthermore, intellectuals of the eighteenth century would criticize both Catholics and Protestants as "superstitious" (Cameron 2005, 145). Thus, the term itself became less about a belief in the supernatural and more about a position against the science, secularism, and rationality of the Enlightenment.[4] In other words, to be superstitious meant to inhabit the position of Other to Christian piety, scientific reason, and "post-Enlightenment logical positivism" (Magliocco 2012, 7; Elzinga 2010); subjects were designated as an "unreasonable faith in arbitrary manifestations of the divine" (Cameron 2005, 40; see also Habermas 1984, 146).[5] Belief in and the practice of superstition and magic persisted into the nineteenth century, but the staying power of that belief has been understood as particularly entrenched in rural Italy, stemming from poverty, poor health care, and sanitation and the manipulation of the supernatural as a "protective strategy" (Magliocco 2004, 156; Eleta 1997). That rural Italians were likely to be imagined as primitive subjects of enchantment was an extension of the already deeply entrenched notion of rural and Southern Italians as uneducated and less modern than urban and Northern Italians. Focusing on how Italy's South has been imagined by Northern Italians and Europe, Nelson Moe (2002) suggests it "was figured as a reservoir of customs and traditions that the modernizing nation was gradually eliminating and for which the middle classes often felt a certain nostalgia" (188). What has resulted, Moe suggests, is an "imaginative displacement" in which modern subjects might work through their ambivalence toward folklore and superstition, among other apparently uncivilized vestiges of pagan and feudal Europe: crime, savagery, tradition, exotica, ignorance (Moe 2002, 51; 188).

Today, the Catholic Church still prohibits forms of divination—fortune telling, astrology, clairvoyance—as "demonic," and it outlaws magic and sorcery or "attempts to tame occult powers," which include the wearing of talismans or charms (Bader, Baker, and Molle 2012, 709). Even "scientific researchers" of religion have found that belief in the paranormal coexists with moderate levels of religiosity in Italy; they frame the trend using market logics: "A tendency for religious consumers to 'diversify' their supernatural assets further compels religious believers to hold paranormal beliefs" (Bader et al. 2012, 707).[6] Ethnologists, such as Cecilia Gatto Trocchi, have shown that supernatural ritual practices have been folded into alternative religions in the late twentieth century and have become widespread in urban areas in Italy (Magliocco 2004, 156; Trocchi 1998; Marchisio and Pisati 1999). Trocchi characterizes contemporary Italy as a space of "spiritual nomadism" (*nomadismo spirituale*), in which people engage in a variety of spiritual practices and where certain mystical beliefs have become normative: over a third of the country believes in reincarnation and extrasensory perception (Trocchi 1998, 11). Trocchi argues that this new renaissance of mysticism, the supernatural, and the occult derives from the conditions of late modern life in Italy: urban isolation, a moral crisis of value, and the loss of the family and basic institutions. The newly popular practice of magic, witchcraft, and occultism in Italy have transitioned into modernity insofar as they manipulate the material world and derive from the idea that the distinction between self and Other can be made arbitrarily (Trocchi 1998,130–131). Here, Trocchi implies that there is nothing intrinsic to Otherness, subordinated status, and even dehumanization, but rather might apply to groups and people based on unique cultural and historical conditions.

The terrain between superstition, science, and religion has long been intertwined with power grabs, mass shifts in governance, and the production of particular forms of knowledge. We also have record that the skeptics, in their knowledge of and masterful manipulation of objects, arise as key figures when the stakes of these shifts are high. Graham Jones (2010) details the way in which modern illusionism became an unlikely but powerful mode of sustaining French empire in his insightful exploration of the 1856 mission to Algeria by "nineteenth-century avatar of magical modernism" Jean-Eugène Robert-Houdin (69). Illusionists, Jones tells us, were seen not only as Western and modern but also as "a scientific popularizer and a debunker of superstitions" (69). Modern tricks and entertainment magic, like Robert-Houdin's, stage the performer's clever and calculated manipulation of the world, while the audience consumes his agentive craft, not supernatural power (69). In this sense, CICAP members are analogous to "paranormal investigators," only their "tricks" involve staging experiments in order to debunk the supernatural. Like Robert-Houdin, they reserve certain kinds of interventions as public performance.

When I visited the Paduan chapter of CICAP in July 2011, I was told of an experiment surrounding a man who said he could feel running water deep below the earth's surface and, in turn, locate where wells should be drawn. The experiment had taken inspiration from one conducted by the French equivalent of CICAP, in 2007 (Albini 2007). The practice is also known as "radiestesia," or water divining, which comes from the words "ray" ("radius") and "sensation" ("aesthesis"). In water divination, which dates to Egypt, Hebrew, Chinese, and Greco-Roman societies, practitioners use instruments such as buckets or pendulums, tools that help them sense the subtle vibrations of running water.[7] In this double-blind experiment, CICAP member Giacomo Lessidri explained that volunteers set up three thick metal tubes in a tent. The experimenter, who, I was told, did not have knowledge himself of which tube had water, conducted several trials asking the water-senser to locate which tube had water, which the experimenter then checked with the outside objective actor who had recorded which tube was turned on for each trial. In front of an audience, the water-senser took time to feel and hover his hands over the tubes before issuing his response. The water diviner's results, Giacamo told me, failed to surpass the statistical likelihood of guessing randomly among the tubes. Like Robert-Houdin's illusionism in Algeria, CICAP's experiment represented a "successful instance of the modern disenchantment of primitive superstition" (Jones 2010, 71). Just as such events endorsed the apparent "cognitive superiority" of the French, so too did CICAP's employment of the experimental paradigm reaffirm the superiority of empiricist epistemologies.

Figure 2.2. A participant in "A Day against Superstition" breaks a mirror, June 2011. Photo by author.

Supplanting Salt with Stats

For CICAP's self-proclaimed soldiers against irrationality, their demonstration in Vicenza's expansive piazza on their Day against Superstition appeared fairly modest. Volunteers assembled a small tent and a table, arrayed pamphlets, set up a smaller wooden table and hoisted a large, ten-foot ladder. In Italy, Friday the 17th is associated with bad luck and malign possibilities, much like the American notion of Friday the 13th. So the Day against Superstition was strategically selected and planned to begin at 5:17 p.m.—17:17 using Italy's twenty-four-hour clock. Precisely at 17:17 of Friday the 17th, one of the CICAP volunteers smashed a mirror with a small hammer, inaugurating the first antisuperstitious ritual.

In addition to mirrors and black cats was an elderly man dressed all in black with a large, broad-brimmed hat. His face had clear, dark eyes and sharp features, and it seemed as if he had leapt into the piazza from a seventeenth-century painting; his angular yet smooth visage was both uncannily familiar and strange. He was dressed as a Neapolitan folkloric masculine figure, literally "the projector" (*jettatore*), meaning one who brings or casts bad luck (*jettatura* or *jetta*) onto others, similar to the evil eye (*malocchio*) but its own category (de Ceglia 2011, 76). One of the volunteers told me proudly, "All day nobody has even approached him!"

As I approached the skeptics' table on the Day against Superstition, I was invited to complete a task designed to undermine superstitious beliefs in the efficacy of particular misfortunate practices. Marco, a trim and bespectacled man in his late thirties, explained how it would work. He presented me with four colored dice and asked me to guess which number I would roll for each one. My guesses were recorded on a large pad of paper set up on an easel, and I then rolled the four dice. My guessing accuracy was calculated at 50 percent, which, as Marco was quick to point out, was already at the proper level of statistical probability. He then invited me to engage in one of the activities they brought along that would ensure bad luck. I had some options: grabbing a handful of salt, sitting closer to the *jettatore*, standing underneath a ladder, or breaking a mirror. There is an Italian proverb about salt: "Drop salt, catch bad luck" (*A versare sale, porta male*). Given my options, I went with the salt. Then, Marco had me guess how I would roll the dice a second time and recorded my answers on a large notepad with each participant's scores. I then rolled them and my accuracy level was precisely the same as before.

Figure 2.3. CICAP's main table featuring the tally of dice rolls at CICAP's "A Day against Superstition," June 2011. Photo by author.

"See? The salt didn't bring you bad luck! It had absolutely no effect on you!" he declared victoriously. In his theory, the salt would have brought me misfortune and made my guesses about the dice roll less accurate. In order to untangle these logics, I must parse the workings and hidden assumptions of probability in relation to superstitious beliefs.

In order to combat my alleged view that throwing salt, a ritualistic practice, would stall events or practices in the world, he provided dice, an alternative material object, and the infrastructure of logic, statistics, and mathematics to nullify my belief in the supernatural power of salt. Both are models of reasoning that rely on particular knowledge relations between human, other human, object, and world: while one is governed by a causative narrative of ritual saliency safeguarding against misfortune, the other engages the objects (the four die) to nullify temporal relations between the present moment and future actions. The array of superstitious practices—mirror shattering, salt throwing, under-the-ladder standing—are animated by underlying kinds of knowledge involving an invisible, sacred, and often future-oriented realm. They are believed to act on the world in a preventative way, by forestalling negative events or happenings. The dice game, by contrast, operates according to scientific methodology by, first and foremost, having an operationalized definition of misfortune in which the ostensible "misfortune" means less accurate guesses. In one CICAP publication on the matter, the authors describe superstition as an error in evaluating evidence, in which an actor misinterprets "proof of the existence of a cause-effect

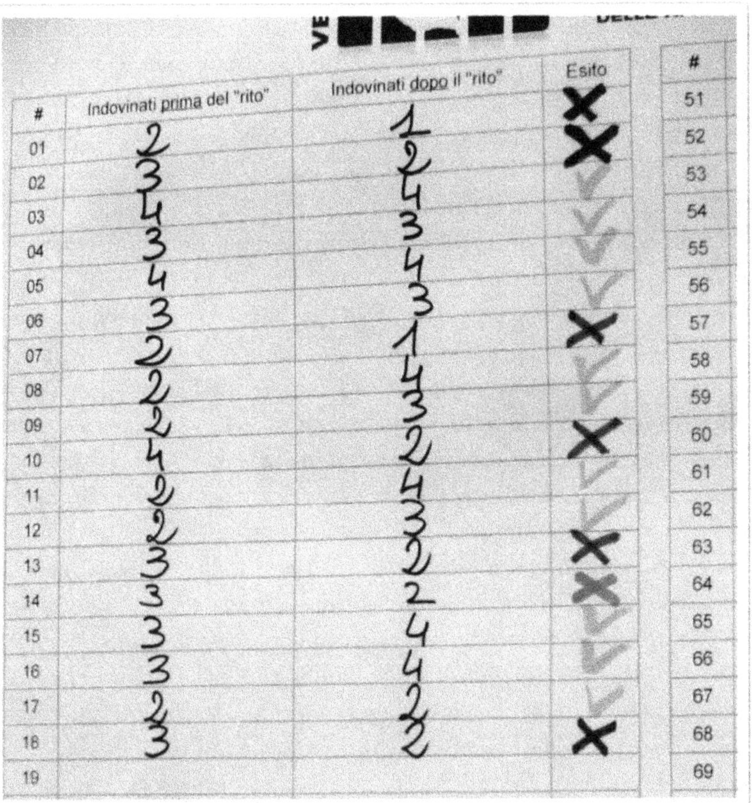

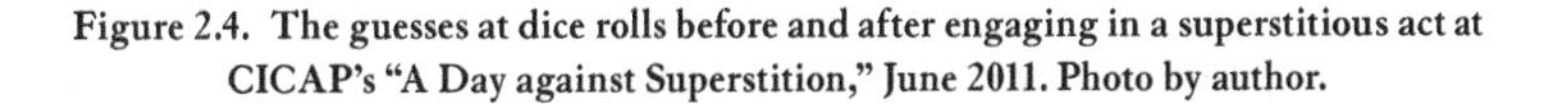

Figure 2.4. The guesses at dice rolls before and after engaging in a superstitious act at CICAP's "A Day against Superstition," June 2011. Photo by author.

relationship" (Zocchi 2011). Within this order, the possible negative event in the world means exclusively fewer accurate guesses of one's subsequent throw of the dice.

The probability game, however, misses the human-object relations of superstition in a few important ways. For starters, in the realm of superstition, the use of material objects and controlled action to regulate the unseen possibility of misfortune does not necessarily have a time limit. A few weeks later, I had come down with the flu in the middle of the summer. One of my friends told me it was because I had not heeded the salt. In the early 1980s, David Marks and Richard Kamman wrote on a concept called "oddmatch" in which evidence that sustains a person's intuitive hunch is remembered, whereas the absence of reinforcing evidence is discounted (Brewer 1992, 123). If I throw salt and do become ill, this reinforces my belief in the power of the salt, whereas had no unfortunate happened, I may have forgotten about the salt. Yet the logics of superstition, I would suggest, may resurface later, remembered as the causal event for a future misfortune, depending on what unfortunate events arises and the meaning ascribed to the failure to fulfill any other protective rituals.

On another level, superstition is more about staving off punishment than producing desirable outcomes in the future; Ernesto De Martino (2004) had it right in calling them "protective rituals" (159). For the believer in the danger of mirror shattering, it is actually a ritual of nondoing, the human act of avoidance, which creates meaning: in exchange for this inactivity, one receives protection and safety. Marco's dice game was still a kind of knowledge practice intertwined with a desire for security, an affect of momentary but soothing certainty. By supplanting salt with statistics, it actually had similar end results: to produce assurance and the affective pleasure of security, the possibility of controlling, if momentarily, the chaos of life. If we consider this particular kind of feeling as immaterial and affective effects, or as "existential goods," after Jackson (2005), then we might conclude, as he does, that "it is precisely this ambiguity that makes it impossible to reduce intersubjective reason to a form of logico-mathematical reason . . . the logic of intersubjectivity never escapes the impress and imprecision of our lived relationship with others" (43). If, then, probability works to persuade superstitious enthusiasts of their beliefs, it may be by replacing one ritual with another and keeping the affective effect intact. Still, CICAP's probability game was very much about staging a hierarchy of knowledge and worldviews: the skeptics and their supporters could publicly renew their vows to mathematical reason. In prior centuries, believers in superstition and magic could be tried and executed, and later, derided as "abject" or "lower class." In the twenty-first century, the superstitious become the targets of a game of chance in the town square. Probability became a compensatory locus of superstitious ritual for CICAP.

But perhaps we should more seriously consider Catholic ontologies as an even less visible framing which haunts the stakes of contemporary superstition in Italy, an ethos of knowing not easily unseated, a reason why probability had only a slim chance to truly transform belief. In the mid-nineteenth century, French writer Ernest Renan declared of Neapolitans, "This people understands nothing but the flesh, matter" (Moe 2002, 74). Superstitious practices mediate between sense, object, and phenomena, and thus also the material, temporal, and symbolic associations through which they are transfigured. And yet this dovetails with Christian origins: Western notions of belief, in fact, "are so influenced by Christian notions of faith that for many thinkers the two are essentially indistinguishable" (Magliocco 2012, 10; Hufford 1995). But here, we must consider not only Christian belief but the difference between Catholics and Protestants in the belief that the Eucharist is or represents Christ's body; an epistemological divide, of course, that is itself drenched with flesh and blood (Orsi 2007). Victor and Edith Turner (1978) called it a "carefully learned theology of incarnation" (71); Donna Haraway (1988) has suggested that the Catholic belief in the "trans-substantiated" flesh of Christ in the host suggests a greater malleability between the material and the semiotic that runs counter to Protestantism (15–16). Robert Orsi (2007), similarly, names this Catholic ethos "the reality of presence," a way of thinking about how belief in the actual body of Christ in the bread and wine represents an essential ontological design of modern Catholicism. He calls for a "radical empiricism of the visible and invisible real," which recalls the value of an intersubjective, what he theorizes as "abundant events." In this sense, the superstitious practitioner might even view the dice as magical objects, acting on the world and the somatic by counteracting the misfortune of the previous act. If Catholic ontology and "the reality of presence" infuse the material world with symbolic resonance, it may help explain why certain superstitions take hold on an epistemic level. In other words, it offers a theory on superstitious belief that is about the roots of knowledge and its form, rather than paganism or folklore.

Precarious Epistemologies

In the context of a massive epistemological shift, where prediction—financial, scientific, neurological, genetic, geological—is both more precise than ever and yet more risky, and computer algorithms know us better than we know ourselves, it is, perhaps, no wonder that forms of skepticism arise in Italian life. Despite their apparent differences, both the paranormal enthusiasts and the skeptics might be best understood in relation to a growing epistemological crisis in Italy: a loss of faith in witnessing and testimony, sensory knowledge and experience, material validations of reality vis-à-vis quantitative data, technological productions of reality. This condition might be more acute in Italy because of its history of both political skepticism and cynicism and because of the contemporary conditions in the late twentieth century that brought the unraveling or uncertainty to civil society: the welfare state, employment regimes, the family, the population crisis, and more.

In *The Philosophy of Berlusconi*, Gianluca Solla characterizes the late twentieth-century Italy as form of absurd power, what he calls "ubuesque," referring to Ubu the King in an 1989 Alfred Jarry absurdist play *Ubu Roi*, and from which we might more simply term absurd precarity. For Solla, this kind of millennial rule in Italy–of which Berlusconi is paradigmatic but not its singular head—is not merely a new term for an irrational form of power. Rather, the fact that power seems to consistently demonstrate itself publicly is abject and ridiculous, yet this technique is essential to the "limitless and arbitrary" nature of this kind of power regime (Solla 2011, 129). It requires the presence of a manifestly comic figure who will be "spectacularly discredited" but whose power is never exclusively bound to this single individual such that everyday power relations can appear, in turn, impersonal. The precariat's ruler, therefore, stages and constantly performs a ridiculous "crisis," and all the while, state power becomes increasingly consolidated. Also central to this form of power is that "fictions [are] elevated to reality"—or "deceit and falsification become essential" at every level of society (Solla 2011, 148). In short, there is a new collective perception of reality in which falsification can no longer be distinguished from truth, and fake evidence may simply be more desirable (Solla 2011, 148).

Solla is interested in laying bare the architecture of a different kind of power, but his insights might help further contextualize the rise of the skeptics, who are radical empiricists and paranormal enthusiasts. The "crisis" endangering the emergence of Italy's scientific skeptics is not only the fraught socioeconomic and political "crisis" but also a crisis of another sort: risk surrounding the underlying veracity and tenability of an empiricist and rationalist epistemology. In this sense, the skeptics' seemingly ardent insistence and public enactments of scientific empiricism might be seen as reactionary or compensatory. In other ways, scientific objectivity obscures an underlying desire to erase doubt. Ella Butler (2010), for example, examines the ways in which the United States' Creation Museum is "mobilized by an epistemological force granted to doubt," specifically, "the problem of Biblical doubt" (236). Rowan Wilken (2011) takes up the issues of post-9/11 security culture in the United States and suggests the crime show *Numb3rs*, in which a mathematical genius helps the FBI solve cases, stages scientific rationalism as a solution to invisible menaces and risks to the social order and state control (203). Here, adopting Agamben's (2002) notion of the management of disorder not control, Wilken sheds light on the fact that "the irony is that scientific rationalism (reason) is mobilized to manage an 'ordinary' made extraordinary by the performative normalization of exceptionalism (or what might be termed a state-imposed enterprise of unreason)" (205).

CICAP members insist on grounding truth in quantifiable and valid empirical claims, in contradistinction to the highly malleable fictions of its absurd and uncertain governing and neoliberal economic regime. To the skeptics, then, they lump together witches, UFOs, holistic medicine, and saint worship, because any form of experience that relies on faith rather than scientific means of truth becomes suspect. They are radical empiricists, devotees of the material validation of truth. Here I do not intend to overstate the significance of their interests or assume that a supernatural phenomenon is a "dependent variable" for some other social fact, as Robert Shanafelt (2004) rightly warns against (327).

But why publicly reaffirm the backward or pathological nature of those inclined toward otherworldly enchantment? Along these lines, CICAP's "Day against Superstition" represented a curiously passionate disavowal: their overenthusiastic and dire assertion of the scientific and rational regime suggests that the apparatus for making valid empirical claims seems suddenly fragile, in ways age-old yet surprisingly new. And precisely here is where, I believe, their urgency to debunk such false truths might stem from a deeper desire to correct and compensate for the seemingly endless array and proliferation of disinformation of Italy's present historical and social crisis. The Day against Superstition, we might say, created magic to combat magic: an illusion of control and logic, a willful attempt to produce stability and coherence. But in doing so, CICAP plays into an age-old scapegoat, the foil of superstition, as a site of mediating much deeper epistemological uncertainties.

Chapter 3

The Rise of Algorithm Populism

The Italozombie is the Country's metastasis and freedom of information is the cure.

—*Il Blog di Beppe Grillo*

Monsters involve all kinds of doubling.

—Elizabeth Grosz

Today, that magic—that bolt of lightning—takes the form of the algorithm. And the algorithm encourages capital to follow its mortal fantasy like never before.

—Stefano Harney, *The Algorithms of Capital: Accelerationism, Consciousness Machines and the Autonomy of the Commune*

In May 2017, then Democratic Party leader and former prime minister Matteo Renzi delivered a scathing speech about the shifting position of the Five Star Movement (*Movimento Cinque Stelle*, M5S)on the European Union: "[The Five Star Movement] has this attitude of having different positions based on whatever is the current trend.[1] Look, this tendency is very strong in the 'algorithm party.' They follow whatever has already happened, in some sense, predetermined, as if it were a Google search, that which gets predetermined by other users' desires. This is not my idea of democracy." ("Renzi: M5S Partito-Algoritmo" 2017).[2]

He referred then to Berlusconi's party Forza Italia as the "business party" and pleaded with listeners to choose his party, the Democratic Party (Partito Democratico; PD), as the only true party. On the one hand, Renzi's "algorithm party" referred to the Five Star Movement's use of algorithms in its digital platform Rousseau, introduced in 2016 by the Casaleggio Association. On the other hand, his framing also implied that the party as a whole worked like an algorithm: automated programming that mined citizens' data for popular trends. Yet Renzi's skepticism apparently found no traction. In the March 2018 elections, the Five Star Movement had a spectacular victory: it earned the majority of votes with 32.22 percent; the League (*Lega*) was not far behind with 17.69 percent, and there were six-point declines for both the Democratic Party and Berlusconi's Forza Italia ("Italian Elections 2018" 2018).[3] The Five Star Movement then formed a coalition government with the right-wing party, the League, to form the first populist government to exclusively lead a Western European country (Alcaro and Tocci 2018, 1; D'Alimonte 2019). In June 2018, the government coalition began its rule with Deputy Prime Ministers Luigi Di Maio (M5S) and Matteo Salvini (*Lega*), along with Prime Minister and former university professor Giuseppe Conte.[4] The Five Star Movement has been called a protest movement, populist, antipolitical, and even anarchist (Biorcio and Natale 2013; Bobba and McDonnell 2016; Caruso 2016). Still, the "algorithm party" had succeeded in ways that amazed the world.

Examined broadly, the Five Star Movement steadily gained political ground and attracted citizens based on a shared disenchantment with mediatized politics, precarious labor policies, and government corruption since its founding in 2009, and built from a grassroots social movement led by Beppe Grillo. Grillo has secured his image as a political outsider and purposeful humorist whose antiestablishment practice makes him the antithesis of media guru, corporate player, and accidental comedian Silvio Berlusconi. By 2012, the Five Star Movement garnered more than 10 percent of the vote in local elections, largely centered in the center-north of Italy, and five mayors and 150 members were voted to local councils (Bordignon and Ceccarini 2013, 5). In the general elections of February 2013, Grillo's Five Star Movement received an astonishing one third of national votes.[5] Political analysts Riccardo Alcaro and Nathalie Tocci (2018) argue the Five Star Movement and the League share populist principles in "fighting the 'establishment,' broadly construed as mainstream parties, bankers, Eurocrats and intellectuals and the liberal, or open political agenda this establishment supports: open borders, multiculturalism and European integration"(1).[6]

The Five Star Movement's success is both foreseeable and surprising. How did Italy shift from the glossy prepackaged world of Berlusconi toward a grassroots, Internet-driven and algorithmic political movement? Why did many Italians fail to trust technocrats like former prime ministers Mario Monti and Enrico Letta yet put faith in people who have been called "a self-appointed elite of web-savvy technocrats" (D'Alimonte 2019, 120)? And why was Italy's populist government's techno-utopian vision also mingled with an enchanted discourse of zombies and ghosts, as well as conspiracy theories and antiscience rhetoric? The story of the Five Star Movement is complex and ironic: odd pathways from media-doubting and techno-phobic citizens who then made media and technology their political foundation; winding roads from grassroots and antiestablishment views to alliances with right-wing political strongholds (Fella and Ruzza 2013; Caruso 2017; Mosca and Tronconi 2019). As Italian political scientist Roberto D'Alimonte (2019) puts it: "In the case of M5S, the web is not just a form of communication and mobilization—the web *is* the movement" (119). The political masterminding of TV gave rise to the political masterminding of the Internet. As the "algorithm party," the Five Star Movement is globally unprecedented in its sophisticated use of citizen-subject information, where the Internet replaced the television as a central political mode of communication and where algorithmic data analysis replaced TV network censorship and tight programming control (Navarria 2019). It is thus key to see the Five Star Movement as a new and innovative political form that grew in relation to and in reaction against material information structures like politicized television and media disinformation of Berlusconian Italy.

The algorithm party's big win in the 2018 elections is one cultural example of a massive shift toward algorithmic ways of knowing in the late twentieth and early twenty-first centuries. Theorist of "algorithmic culture" Ted Striphas defines an algorithm as "a set of mathematical procedures whose purpose is to expose some truth or tendency about the world" (Striphas 2015, 404). Algorithms are computational processes, but they have become saturated in our networks of knowledge because of "Big Data" on the one hand and the uptick in forecasting data analysis, or what is also known as "predictive analytics," on the other. Big Data distinguishes information that comes from massive, hybridized forms of available information: from Google maps to online shopping to mapping technologies to public health and environmental risk (Chandler 2015, 835; Nafus 2018, 372). Scholars, too, have been racing to theorize the implications of algorithms in our world in every sphere of life: "algorithmic culture" (Striphas 2015), "the datafication of everyday life (Chandler 2015, 636), the "algorithmic self" (Pasquale 2015, 32), "algorithmic identit[ies]" (Cheney-Lippold 2011, 165), "algorithmic forms of sociality" (Wilf 2013, 719), "algorithmic regulation" (McQuillan 2015, 566), "algorithmic manipulation" (van Dijck 2014, 200), and "algorithmic assemblages" (Lowrie 2018, 350).

For these theorists, the new paradigm means that algorithms undergird and remake social and moral life in ways that are largely unseen or at least obscured, and almost all agree that it results in the concentration of power in fewer hands. The "age of the algorithm" is becoming a totalizing force that shapes, as some scholars have put it, "new geopolitical and geocultural configurations of information, labour and economy" (Rossiter and Zehle 2014, 126). From car-share services like Uber to online sales giants like Amazon, algorithms represent a hypercommodification of individual consumer data (Ekbia and Nardi 2018, 365). Adopting the term "algorithmic culture," Ted Striphas (2015) argues that algorithms have changed not only culture but also its ontology, or as he puts it, "how the category culture has long been practiced, experienced and understood" (396).[7] The material infrastructure of Five Star's algorithmic digital interface and its emergence through an Internet blog shape not only the democratic practices but also political culture, broadly construed.

To further delve into answering the questions of why and how the Five Star Movement algorithms, populism, and even antiscience intersect and co-

Figure 3.1. Beppe Grillo in Frascati, Italy, at Five Star Movement rally, 2009. Photo by Claudio Bisegni.

ignite each other, I will explore the roots of the founder of the Five Star Movement, Beppe Grillo, and make sense of why tales of zombies and ghosts have more in common with media critique and technological innovation than may be immediately evident. Finally, the chapter takes a close look at the creation of Rousseau, the algorithmic program of the Five Star Movement and the party's populist fantasy of Internet direct democracy. The emergence of Italy's "algorithm party" was both a reaction to and development from Berlusconian politics, which also, in a darker sense, made Five Star enthusiasts somewhat less skeptical of how digital democracy can undermine truth and exploit participants in covert ways.

Zombie Imaginaries

In November of 2011, Beppe Grillo—a social critic and Italy's comedian-cum-politician who leads the populist Five Star Movement (*Movimento Cinque Stelle*)—cracked a Berlusconi joke: "Berlusconi is an old zombie. He was already politically dead in 2008" (Grillo Blog, November 9, 2011). It gets a laugh because despite Berlusconi's numerous legal indictments, corruption and prostitution scandals, and political gaffes, he's outlived several political deaths since his election to prime minister in 1994. Grillo also teased Berlusconi for his infamous cronyism, remarking, "We always see solidarity between zombies" (Grillo Blog, December 26, 2011). He regularly refers to former prime minister Mario Monti as "Rigor Montis," the political parties of Italy's Second Republic as "zombies," and Democratic candidate Walter Bersani as "the dead man who speaks" (Bordignon and Ceccarini 2013, 10, 20; Grillo Blog, February 27, 2013). Grillo referred to Prime Minister Letta's government as the "Monster Club" (*Il Club dei mostri*) and called it "mythological, a fantastic animal, with many heads but only one brain with two separated hemispheres: the right one is Berlusconi which is already predetermined, and the left one is international finance ready to squeeze Italy" (Grillo Blog, April 28, 2013). To his mind, Italy's political system is like a delusional ghost: "Our weed-ridden politics are like the spirit of a dead person that doesn't know he's dead" (Grillo Blog, May 6, 2010). One of his 2013 political posters features a fat zombie with a mohawk, whose protruding belly shows a last meal of various opposing politicians, and which has the tagline "Politicians are scared of us, you know why? Because we eat them!"

Grillo's message to the Italian electorate is clear: it's a zombie-eat-zombie world.

Certainly Grillo's flair for humor and satire echoes and challenges the Berlusconian status quo. Indeed, it is hardly surprising that Grillo has great jokes on government corruption: his quip exposing former prime minister Benito Craxi's corruption is legendary in Italy; his biting one-liners, popular since the 1980s, on exploitative governmental and corporate power have made him trustable and likable among many Italians, and his humor is surely a critical feature of his party's appeal. Along such lines, Roberto Biorcio and Paolo Natale (2013) suggest Grillo has cleverly marshaled "his original vocation of satirist and radical critic of political parties" (16). Fabio Bordignon and Luigi Ceccarini (2013), describing Grillo as part "anti-political sentiment" and part "political entrepreneur," attribute his popularity to Italians' "uncertainty about the future and demand for protection clash[ing] with the anti-crisis measures, stoking the anti-party sentiment" (2). So, at first glance, we have a familiar story: Grillo, like other funny politicians around the globe (Baym and Jones 2012), has deployed humor to bolster his antiestablishment authenticity. Across Europe and the United States, we have seen politicians and political activists intertwine humor and satire as successful strategies (Haugerud 2012; Molé 2013). In an age of increasingly orchestrated and consolidated social and political messages, political humor has increasingly become the necessary weapon of political dissent and the makings of populist movements (Klumbytė 2014; Boyer and Yurchak 2010).

Though Grillo's appeal may have been a result of his being a political outsider when standard politics looks dismal, this explanation needs to account for a salient social fact: Grillo's regular use of zombies and ghosts. Grillo's supernatural humor deserves more credit in propelling his movement than has otherwise been recognized. His otherworldly humor does not just speak to brewing cultural and economic anxieties but rather represents a kind of wink to citizens: he, too, recognizes the profound and ghastly irrationality of contemporary Italian political life and, more importantly, the unreliability of television and print media as information sources. This wink communicates what many Italians secretly long to hear: you are not insane. Just as the zombie becomes the figure of Italy's disinformation society, it also enables a kind of exorcism. Grillo's promise to exorcise Berlusconi's media demons makes room for his own unique party brand where algorithms become, ironically, the new hidden demons of digital populism.

The Rise of the Digital Populist

Before we return to analyze the significance of his enchantment tropes, let us first take a deeper dive into Grillo's history as a performer and social critic, recalling that the Five Star Movement was built on Grillo's grassroots following. Grillo's aim was not to be elected himself but to be part of a party that engaged in transparent democratic practice.[8] Over the course of three decades, Grillo has earned his political chops in unconventional ways. Artistic director and television host Pippo Baudo helped Grillo get his first television show, *Videobox*, in 1977 (Galeazzi 2013). Starring in several variety shows such as *I'll Give You America*, Grillo joined forces with famous producer and infamous satirist Antonio Ricci, who is known for bringing the first news parody show to Italian television in 1988, *The News Is Creeping* (*Striscia la Notizia*) (Molé 2013a). In 1986, on the Ricci-produced program *Fantastico*, Grillo became infamous for joking about the then socialist prime minister Benito Craxi: "If the Chinese are *all* Socialists, then who do they steal from?" Grillo's whistleblowing joke was prescient. In 1992 the "Tangentopoli" investigation led to Craxi's indictment on charges of corruption and bribery; Craxi eventually fled to Tunis to escape imprisonment. The Craxi joke also had significant negative consequences for Grillo: he was chastised by Craxi himself, lost Baudo's support, and got his airtime reduced by the Radiotelevisione Italiana (RAI) network directors, and eventually his show, *The Beppe Grillo Show*, was canceled. The event would represent a personal trauma for Grillo from Italy's censoring television media regime, and perhaps helped motivate his later turn to and reliance on his blog.

After his ousting from network television, he did stand-up and satire through the 1990s until the creation of his blog in 2005. Receiving recognition and awards from various news outlets including *Time* and *Forbes*, Grillo's highly acclaimed blog exposes corporate corruption and disseminates critiques of neoliberal capitalism (Bordignon and Ceccarini 2013, 3). It is also a forum for breaking stories on corruption, environmental issues, and health care; he was especially famous for anticipating Parmalat's corporate abuses, for which he later served as a witness for the prosecution. His 2007 book, *Modern Slaves: Precarity in Italy of Wonders* (Schiavi Moderni: Il Precario nell' Italia della Meraviglie), decried the economic policies of the 1990s and 2000s that moved Italy's highly protectionist labor regime toward a market-oriented system, a problem disproportionately affecting Italy's youth. Because Berlusconi ushered in these market-oriented labor policies, Grillo's public contestation secured his image as against, or counter to, the prevailing political and information regime.

In September 2007 Grillo organized "V-Day" or F—Off Day (Vaffa' Day), a public rally against journalists and politicians, the latter whom he called the "real unlawful people" (Bordignon and Ceccarini 2013, 7). Eugenio Scalfari, editor of the Italian magazine *L'espresso*, called it Grillo's "barbaric invasion" and Grillo himself an "anarchist and individualist of our people" (Biorcio and Natale 2013, 89). Grillo responded to criticism of the first V-Day and the public insistence on whether or not he was creating a political party: "Parties are encrustations of democracy. You need to give space to citizens. . . . We live in party-ocracy not in a democracy" (Biorcio and Natale 2013, 90). Grillo organized a second V-Day on April 25, 2008, against what he had dubbed Italy's "caste of journalists," and collected signatures for referendums against public financing of editorships (Biorcio and Natale 2013, 80). The Five Star Movement became an official party in October 2009 although Grillo, promising to rebuild Italy "from the bottom up," required that his constituents *not* be members of existing political parties (Bordignon and Ceccarini 2013, 8). He extended his populist, grassroots, and outsider vision to his party: its mandatory political-outsider membership, its nonhierarchical structure, and its engagement in direct voting within the group. As a whole, the platform's "rejection of the political class" has been a distinguishing characteristic of contemporary Italian populism (Agnew and Shin 2017, 917). Their main platform included a stance against the privatization of water and the politicization of the media, and in favor of the environment and increasing public transportation.

Grillo has long been especially aggressive about media and news corruption, or, as he calls it, "state-assisted information, freely bought by politicians and financiers" (Grillo Blog, April 9, 2007). Popular cynicism with news media stems in part from Italy's midcentury system of "sharing out" (*lottizzazione*) political power and dedicating one news network per political party. This system was replaced with media tycoon Berlusconi's 1980s privatization, which produced a two-tiered system between private and public channels, with even more flagrant political ownership and control of the media (Briziarelli 2011, 13). During the 1990s and 2000s, Berlusconi represented the era of "mediatization of politics," wherein theatrical and media-ready politics—mostly televised and in print media—became the norm across many Western democracies (Campus 2010, 227; Boyer and Yurchak 2010). Moreover, Berlusconi also used a method of overt censorship to silence political satirists off the air, and shows that criticized him or his party struggled to get airtime (Edwards 2005). Grillo's movement and the Five Star party were clear responses to decades of cynicism and distrust over the power-saturated television networks and print media.

Sofia Cingolani is a fifty-something Five Star Movement voter who has lived in the United States for ten years but avidly follows Italian politics.[9] She began narrating her shift from the Radical Party (*Partito Radicale*) to the Five Star Movement around 2009, which was also when Berlusconi was embroiled in scandal: "Europe forced the Berlusconi scandal. It was imposed by Europe, you know that, right? Berlusconi has to go because of the scandal. Whom did the put? They put in Monti but why? To drain all the money from the poor to the rich. If you go to see what Malvezzi says—he's one of the economists who says there is no crisis. The 'crisis' was something written up at a little table somewhere in order to drain money from the middle class to the rich, to the banks." Sofia's understanding of politics is informed by the view of Italian politics that is shared by many Five Star Movement enthusiasts and guided by a few principles: (1) Italian politics are manipulated externally so politicians and political decisions in Italy benefit non-Italians: power structures like the European Union and largely hidden networks of global finance; (2) television and print media are mere puppets for these political institutions and interests; and (3) you can find true and reliable counterinformation online. To clarify, she referred to the economist Valerio Malvezzi, who is a university professor of political science at the University of Pavia, former deputy for the League and known critic of global finance. Later she discussed the French populist movement known as the "yellow jackets," and said how the movement includes both the Left and the Right, adding, "It's a level of social discontent that traverses the political spectrum; here there's people from the Right and the Left that feel betrayed by politics and go together to the piazzas." She noted that the French movement was hardly covered on French and Italian television: "Media are held by big corporations."

Sofia then explained that this deep-rooted and pernicious financial incentive affects major newspapers like *Corriere della Sera* and *La Repubblica*, which printed the "same made-up stories" and were owned by banks, and television was entirely controlled by banks, hedge funds, and bank-controlled politicians. She referred to Francesco Amodeo's self-published book *The European Matrix* along with his online interviews as revealing the shadow financial network that controls the world—Bilderberg, Trilateral, and the Group of 30 (Gruppo di 30). She added, "You understand the long arm of finance and how far it reaches. . . . You understand that we are really slaves to power which at least for now—there's no escape. For a person like me, we only have the power of our vote. We don't have any other power." The Five Star Movement supported renationalization of banks, which was one of the most important issues to her.

Sofia also recalled her first time at a Grillo event:

> I had already gone to hear Grillo—it was in 2005 or 2004. He came to Rimini to talk about incinerators. Do you know what incinerators are? They call them "waste-to-energy" plants [*termovalorizzatori*]; they are the plants that burn waste and produce all of these nanoparticles that go everywhere. They get inside plant cells and then we eat this stuff since it grows in the plants, and it's very harmful. And Grillo was already talking about this during the years of his show. He was founding the Five Star Movement but it wasn't yet founded officially. It was a popular movement which started from below [*partito dal basso*] with people who went to his shows. When he saw he could fill a twenty-thousand-person arena they said, "Wait a minute," with Casaleggio. Casaleggio is this wonderful person who had a dream. These people were normal people who wanted justice. Politicians are on the corporate payroll: all of them! From the Right to the Left, including the unions, they are all paid by them. That's why they don't follow people's interests. And so they had this idea: there's an opening. Here we have people who think as we do.

Sofia also expressed a contempt for PD that I have heard from many Italian friends and informants, a kind of betrayal that the Left was able to execute the wishes of the wealthy. My friend of over fifteen years, Paduan union activist Helena, assured me that the co-opting of PD was a clever political move, a strategy designed to trick people into believing the motives would benefit the working class even though they undermined them. Sofia said, "I've always been disgusted by PD; I'm like a psychic because I saw the corruption of PD already in Berlinguer's time. I saw PD like a big corrupt freak since his time. Someone like D'Alema with his artisanal shoes that cost 15,000 euros or 10,000 euros? And he was supposed to represent workers, you think? Come on, you can see it." The deep disgust and cynicism with the Italian Left go back decades, and further explain why a party that rejects both the Left and the Right might be desirable. Grillo has criticized politicians' maligned motivation, the idea that they are guided by interests other than those of people, but also implied the idea of a conspiratorial government that manipulates people through the media and the appearance of different parties, when in fact they all are motivated by the same financial groups and interests. Sofia's savvy take on Grillo reflects these same moral principles as well as a hypervigilance regarding the corrupt motives of the existing political system. For voters like Sofia, the populism of the Five Star Movement is driven by a desire for greater truth and transparency not just of its political leaders but also of systems of communication.

Italozombies: The Walking Dead of Televised and News Media

Let us now return to the supernatural discourse Grillo has used in his blog with an eye toward its deeper significance as a critique of the material structures of information and media. In 2010, for example, Grillo reimagined the Italian citizen as the "Italozombie":

> Italians . . . are second generation zombies. . . . Italozombies are united with the traditional zombie in its apparent incapacity to plan or to desire anything. In reality it is a victim of idiotic transitional laws that have been propagated via news and television. When it makes a wrong decision it keeps going at the cost of appearing like an imbecile, until it actually becomes one. A consistent feature of stupid-ified Italozombies is the envy of the person who wants to emulate them. . . . The Italozombie gets aggressive with anyone who tries to explain that it is acting like an Italozombie. . . . It doesn't want to be contradicted nor does it want to discover its true nature. Its daily behavior is media-directed, it doesn't have ideas or opinions. . . . Moreover, it has a temporary memory lasting only a few days and it cannot make any connections between events. The Italozombie is a social predator. . . . Wherever he passes solidarity cannot grow, consciousness cannot be raised, grass grows no more. The Italozombie is a chameleon, it mimics yet-to-be-infected Italians. (Grillo Blog, January 17, 2010)

Consistent with his views on media, Grillo's scary satire imagines a late liberal society gone awry: consumers become mind-numbed cannibals. The zombie becomes a caricature of selfish, apathetic, and vapid citizens made dead by their media consumption and their lack of awareness and compassion. But the narrative also implies hope for Italians in that an infection, an external source, is the cause, suggesting unintentional victimization of the populace, and, therefore, a political possibility that it might be cured. The passage also implies a critique of Italy's most famous media tycoon, Berlusconi, who was at the forefront of secularizing and commodifying privatized knowledge vis-à-vis television and print media in Italy (Agnew and Shin 2017). Berlusconi is information society's zombie master to Grillo's zombie hunter.

In fact, the post ends by mobilizing political action: "The Italozombie is the Country's metastasis and freedom of information is the cure" (Grillo Blog, January 17, 2010). The cancerous media, then, is what erases and decapitates the population: consuming toxic knowledge poisons Italian souls. Still, Grillo also warns citizens that Italozombies "mimic" regular Italians, meaning that there are two groups: visible zombies and invisible zombies who appear human on the surface. Grillo thus hints at a manipulation of consciousness so profound that one might not even know of one's own zombiehood. Citizens, he implies, must scrutinize their own naïve trust in information society in order to root out toxic self-awareness. The notion of hidden corruption also echoes a central political principle of the Five Star Movement.

Narratives about zombies and monsters often emerge in sociopolitical contexts rife with socioeconomic change and widespread uncertainty (Comaroff and Comaroff 2000; Dendle 2007; McNally 2011). The figure of the zombie, the autopiloted human, without motivation or consciousness has also, ironically, become a metaphor for certain algorithmic processes as deadened "roboprocesses," stripped of human agency (Besteman 2019; Gusterson 2019). In Catherine Besteman's (2019) "Afterword" for *Life by Algorithms*, she and coeditor Hugh Gusterson refer to algorithms as roboprocesses, and she argues they result in "zombification," which "degrades critical thinking and the ability to imagine alternatives" and impoverishes interpersonal relationships (15–16). She observes that the figure of the zombie has also returned to social imaginaries in the United States and argues that the algorithmic process drives further zombification as the robo-testing or robot-surveillance drives social structures with little or no human engagement (Besteman 2019, 16). The association between algorithms and zombies is at odds with Grillo's use of the zombie indirectly which persuades citizens to invest in and trust the Internet and algorithm-driven digital platform of the Five Star Movement. For Grillo, the zombie is created by television and print media and the Internet is its only salvation, which may entice his followers to enact a kind of trust in online communication and algorithmic processes.

Then again, we might see zombies as figures precisely because the material structures of information represent cultural paradoxes. Annalee Newitz (2006) examines how nineteenth-century American narratives featuring monsters "embody the contradictions of a culture where making a living often feels like dying" (2). Avery Gordon (2008) suggests that hauntings represent "one way in which abusive systems of power make themselves known and their impacts felt in everyday life" (xvi).[10] The Italozombies of Grillo, who is someone accused of being an anarchist and a demagogue, are figures that help make sense of the cultural paradoxes of knowledge in twenty-first-century Italy, and the way government has co-opted information in order to consolidate power.

Zombies are not the only supernatural stars in Grillo's bag of tricks. In fact, it is only by looking at the full spectrum of enchanted beings that we see that Grillo's intervention is not just about *what* we know but *how* we know. One Five Star Movement political poster shows a large picture of a pig saying, "Every time you call a politician a pig you offend me, even though I never did anything to you. Now, either you don't call them pigs anymore, or you change your politicians." Likening politicians to pigs is old hat, but the twist here is imagining the pig's reasoned response to the joke. The surreal image of the hyperrational pig—an animal more rational than existing politicians—implies that the enchanted reality may be just as coherent as supposedly disenchanted logic. Put differently, it plays on the assumption that the magical talking pig should be nonsensical. In another post, next to an image of a mirror with a glowing-eyed skull in its center, Grillo writes: "Like the witch in *Snow White*, people in our technological society look into the mirror of their clean little society, dominated by capital, and yet they see beauty, extreme beauty" (Grillo Blog, January 28, 2006). Consider, too, one of his famous one-liners: "One Italian is a Latin lover, two Italians make a mess, and three Italians make four political parties." Italian life, he reminds his listeners, does not adhere to basic arithmetic: his jokes insinuate a fundamental break with assumptions about rational order, logic, and scientific reasoning.

One of Grillo's YouTube speeches is set to the scary music of *Poltergeist*, with images of politicians appearing as white-eyed ghosts. "The Italian state," he proclaims, "is infested with political spirits that were with us once upon a time. . . . Holograms of a world that ceased to exist for some time. The real and the perceived overlay each other, melt into each other, become one thing in the minds of Italians" (Grillo Blog, May 6, 2010). Citizens, he implies, are prey to alluring but false representations not because of their individualized insanity but because of a collective and public "melting" or confusion between fact and fiction, the by-product of an incredibly subversive and irrational political system. The enchantment trope has not just been a smart politician cleverly proposing himself, and later the Five Star Movement, as a populist corrective to the status quo, the anti-Berlusconi; it is a masterful rhetorical move that redirects the cynical energy of a population for whom magical thinking, at least within Berlusconian Italy, has become standard political rationale.

The work of supernatural stories, then, is actually epistemological: they are about how knowledge is made, processed, and obstructed. Grillo's zombie figure reaffirms a vision of a split reality within Italy, the danger of something horrible and monstrous appearing authentically alive and human. The zombie's savvy "mimicry" of the uninfected warns Italians that they must discern between the two competing realities: the message arouses vigilance among citizens and makes their already existing anxieties warranted and safe. Whether it was Berlusconi's sex parties or his prosecution for hiring underage prostitutes or his privately owned media conglomeration that he mobilized to get him elected, the age of Berlusconi was riddled with seemingly impossible truths. Moreover, Italy's increasingly outlandish and spectacular public and political culture unfolds almost entirely through media consumption (Molé 2013a). The zombification of the population, in Grillo's vision, results from citizens' mutated knowledge processing and seemingly mutant true information. In a blogpost called "The New Monsters" he writes, "Italian information is decomposing. It's like a journalism of new monsters that are eating our brains. . . . News is the triumph of unburied" (Grillo Blog, April 9, 2007). For Grillo, Italy is haunted by mass-produced false reality propagated through zombified news and television media. Correcting where Italians find information is, at least for Grillo, a high-stakes political objective: people's souls are at stake. The zombie figure, whether it is a person or a set of media institutions, becomes shorthand for a warning to citizens that reality is indeed stranger than fiction; furthermore, citizens must overcome their blind faith in media in order to understand this deceptive duality of the real.

Antiscience Conspiracies

Enchantment discourse stands in for epistemological arguments about the materiality of knowledge production; thus, the supernatural is not necessarily a sign of antiscience. That said, Grillo, and to some extent the Five Star Movement, has also become a figure so skeptical of media and establishment politics that Grillo marshals the same logic toward science: doubting it precisely because it often presents itself as authoritatively true. Grillo has been accused of spreading outrageous pseudoscientific theories like the inexistence of HIV, the carcinogenic prevention of cancer, and the danger of vaccines (Arcovio 2012; Mautino 2012). On his blog he has spread views like "You can die from vaccines" (Grillo Blog, June 8, 2010). Journalist David Allegranti mocked the Five Star Movement for its antiscience views: "Grillo as a threat? Come on! There are deputies who believe in mermaids; there are deputies who believe that the jet streams you see in the sky are chemical weapons!" (Kramer 2015, 45).

In January 2019, the head of the National Health Institute (ISS, Istituto Superiore di Sanità), Walter Ricciardi, explained his late December resignation because of the unscientific positions of Italy's government (Fubini 2019; "Walter Ricciardi" 2019). Ricciardi remarked about Five Star leader Matteo Salvini, "When the Deputy Prime Minister, a father, says that he thinks vaccines are useless and dangerous, this is not only a non-scientific idea. It is anti-scientific" ("Walter Ricciardi" 2019). This ethos of this antiscience rhetoric is often called "denialism" (*negazionismo*), which implies this tendency to deny any universal claim to truth, whether that truth is political, environmental, or medical.[11] Ricciardi also criticized Salvini's unfounded claims that incoming migrants spread disease. On the one hand, the government coalition spreads disinformation about vaccines and xenophobic theories of disease. On the other hand, there is a thread between the foundational principles of media criticism and political cynicism that give rise to scrutiny of all forms of knowledge production, including scientific discourse.

The People's Algorithm

The other vital part of this story is that Grillo's blog was his first collaboration with Gianroberto Casaleggio, who would go on to create the digital platform Rousseau, the algorithmic digital platform from which M5S's policies emerge and constituents make their voices heard (Navarria 2019, 171).[12] Thus we must reexamine Grillo's blog and grassroots organizing as a means toward a very spectacular end: the world's first algorithm party. Casaleggio had started Webegg in the 1990s, a large web consulting firm in Italy, and long envisioned that the web could create new political forms. In his 2001 book, *The Web Is Dead, Long Live the Web* (*Il Web é morto, viva il Web*), he imagined the rule of the people as increasingly greater than single political leaders: "Referenda on topics of national importance will become as routine as reading the papers or the evening news. The interactive leader will then be the new politician" (qtd. in Loucaides 2019, 86).[13] After meeting Grillo in 2004, Casaleggio helped him launch his blog in 2005, which gave Grillo a new and direct way to communicate with his fans that, importantly, sidestepped the Berlusconi-controlled airwaves of television and print media. In fact, it soon became known as the heart of "counterinformation." To citizens immersed in right-wing mediatization soundbites, this was a new pleasure and site of identification (Loucaides 2019, 87). By 2008, Grillo's blog was the number one Internet site in Italy and, globally, in the top ten blogs (Loucaides 2019, 87). Some say that Casaleggio was actually ghost writing Grillo's blog the entire time (Loucaides 2019, 87).

Gianroberto Casaleggio has been called a "cyberutopian" (Loucaides 2019) and "the father of direct democracy" (Barcellona 2016) and Rousseau his "most advanced instrument" (Giua 2016; Natale and Ballatore 2014; Rotondi 2017). It was officially released in 2015, from a prior version in 2013, and offers M5S members the opportunity to discuss law and policy at every level of politics ("M5s, ecco 'Rousseau'" 2015). The platform also allows activism by supplying materials, sharing, e-learning about political structures and procedures, a law-writing feature, and meetup information (Barcellona 2016); Rousseau allows for "a form of legislative crowd-sourcing which can be discussed, amended, and voted on by members online" before moving to Parliament (Navarria 2019, 199-200). In the 2013 election, some Five Star elected officials were actually quite shocked to be elected because the digital "direct democracy platform" had helped put 160 candidates without political experience into office (Loucaides 2019, 82). The site was also created based on Casaleggio's directive to his "digital soldiers" to mimic Google Analytics and Facebook Insights in order to predict and track viral posts as well as adversarial and dissenting information (Loucaides 2019, 92). Rousseau is an enactment of what some algorithm theorists envision: "The ability to directly supervise, intervene into, and operate extended material systems by mathematical fiat—is the basis for a truly epochal metastasis of algorithmic control" (Lowrie 2018, 352). The party's digital platform is central to their radical political agenda as Loris Caruso (2017) puts it: "new forms of digital communication, by enabling unmediated expression of the popular will, make true, unmediated democracy possible" (586). Still, let us recall that Grillo himself had used the illness metaphor of metastasis to mark the scourge of print and television media, not Internet or digital platforms. Yet Rousseau certainly represents a new way to materially mine and produce knowledge as a direct strategy for creating political governance.

Part of how Grillo offsets these concerns is to consistently frame Rousseau as a constantly evolving system, growing in relation to people's voices, and thus fundamentally democratic. Grillo invited members to participate directly with "suggestions, criticism, proposals, and evidence of eventual errors" ("M5s, ecco 'Rousseau'" 2015). In an interview with a Swiss journalist, Grillo said, "You've had direct democracy for 150 years here with referendums. Today there's the Internet and you can have a referendum from home with a click. You can express opinions on everything. My movement wants to give people the means, and occasionally the electronic means, to make decisions about current affairs, to participate in current affairs, to approve participatory budgets, to understand and decide what to do with their money, with their cities, and with their nation" ("Intervista RTS" 2018). Rousseau is not open source, but available and protected only to party members; it also allows members to vote in local and regional elections based on their location and profile (Barcellona 2016). The platform allows constituents to discuss and propose laws, ranging from abolishing police officers from carrying weapons and banning plastic utensils to creating "Fertility Day" to help the country's low birthrates or fighting corruption by abolishing subcontracting

(Buonfiglioli 2016). On the "Law Subscribers" (Lex iscritti) function of Rousseau, and from over 3,280 direct proposals, six would be proposed in Parliament (Buonfiglioli 2016).[14]

Grillo had infamously joked that Rousseau would be a "traitor catcher" (*caccia al traditore*), adding: "We're working on a program, the so-called blockchain with encrypted information. If it's used in politics it could be interesting: you have an algorithm, there are no other intermediaries, if a parliamentary representative votes and hasn't followed the program, they'll be automatically expelled" (Casalini 2016). One Democratic Party (PD) member said Rousseau would be "rolling in his grave," while another said, "Everything damages citizens who face Grillo and Casaleggio Association's decisions every morning about who to expel with its elusive algorithms" (Casalini 2016). Yet the threat of expulsion has not deterred party members from using the platform, especially as it resonates with Grillo's longtime message of anticorruption and antiestablishment. That he frames algorithms as safeguarding party communications is significant as a way to deter skepticism. Moreover, Grillo promoted the transparency and trust in the platform by issuing its use to Casaleggio Association as a nonprofit entity, and the declaration states the association "never had expenses for electoral campaigns" and had "neither received nor given any private or public contribution" ("M5s, il simbolo" 2018).The party also proposed to use algorithms to determine which other party might be most viable for a government coalition. Before the official coalition between Five Star and the League, Luigi Di Maio announced in April 2018 that he would create a scientific committee that would develop an algorithm to analyze the overlapping interests between the Five Star Movement and other political parties (Pucciarelli 2018).

And yet Grillo is sometimes quite skeptical about algorithms, even as his party can derive platforms and agendas based on Rousseau's algorithmic properties and explicitly use algorithms to create alliances. It is here that we find a deeper paradox that combines his deep-rooted skepticism about media and information: "Humanity finds itself facing a new and grave reality: the rise of autonomous weapons. . . . What kind of future looms if human life is left to algorithms? . . . What level of human control is necessary to guarantee both the compatibility with our laws and the acceptance of our values?" ("La Guerra degli Algoritmi" 2018). In this vision, algorithms are framed as highly volatile and dangerous, yet with an implied resolution that "human control" can deter their danger. In another entry, with an image of a humanoid robot with blue gleaming eyes, he also stirs fears: "Now is the era of machine learning. . . . Algorithms learn by themselves" ("Il regno degli Algoritmi" 2018). He describes how two robots, designed by Facebook, started their own language which was not comprehensible to humans, and added: "As it is happening ever more frequently, computers will show us the realization of theory and process, in a way we don't understand. It'll be in that moment that we must begin to trust ourselves and the reign of algorithms will begin." Let us first recognize that Grillo's vision is a totalizing and dystopic version of algorithm culture, not a digital platform like Rousseau which facilitates but does not automate political decisions. Nevertheless, he still asks his constituents to both trust and distrust algorithmic processes while simultaneously ushering in Rousseau as a material instantiation of algorithmic "reign."

Why Rousseau?

The name of the platform, Rousseau, is part of a discursive maneuver to align techno-governance with democratic ideals. On February 9, 2018, shortly before the national elections, Beppe Grillo stood in front of Rousseau's statue in Geneva, high-fiving the marble base he declared, and addressing the philosopher directly: "I'd love to ask you what you think about Casaleggio, about Rousseau, about our direct democracy. . . . You've been our inspiration, our symbol. . . . You hated Voltaire! You made some of the biggest trouble in history! . . . You created the principles of democracy and the French revolution! And we—your small, insignificant disciples—will bring your word to the Italians, we Calvinists! . . . We have to democratize a population that are Italians!" ("Onore alla Democrazia" 2018).

Here, Grillo raises and shakes his own cupped hands together—a classic gesture suggesting something impossible or challenging—then continues: "We can do it, Jean-Jacques! And now Davide Casaleggio has your inheritance and he'll do everything possible to bring democracy to the Italians . . . and we'll come back here with our computers to copy this wonderful city!"

Jean-Jacques Rousseau's Enlightenment philosophies imagined society as a corrupting force to a naturally good-natured and free citizen, for whom the government's role would be to sustain a "social contract" which would stave off vice, corruption, and the psychic ravages of wealth divides and inequality. In analyzing Rousseau's tie to global populist movements, *New Yorker* writer Pankaj Mishra (2016) suggests the philosopher "thrived on his loathing of metropolitan vanity, his distrust of technocrats and of international trade, and his advocacy of traditional mores." With their loathing of Rome's insiders, elites, and intellectuals, especially Berlusconi's much vaunted wealth and lavishness, their Euro-skepticism and anti-immigrant stance, and their ardent environmentalism, it is no wonder that Five Star constituents see themselves in Rousseau's legacy.[15] In a Rousseauian lamentation for moral crisis, Grillo has called Berlusconi "the incarnation of the lack of morals that exists today in Italy" ("Intervista RTS" 2018). Moreover, and unlike other Enlightenment philosophers praising science, Rousseau warned that minds could be "corrupted" by arts and sciences in his 1750 "A Discourse on the Moral Effects of the Arts and Sciences." His anti-intellectual views were largely based on his own exclusion from Parisian intellectuals and urban elites yet on a deeper level inform an undercurrent in his work where victimhood has "incendiary appeal" (Mishra 2016).

His estrangement and distaste for public artifice were particularly evident in Rousseau's "Dissertation on the Origin and Foundation of the Inequality of Mankind," when he wrote, "In the midst of so much philosophy, humanity, and civilization, and of such sublime codes of morality, we have nothing to show for ourselves but a frivolous and deceitful appearance, honor without virtue, reason without wisdom, pleasure without happiness" (Mishra 2016). It is here that Rousseau's words most resonate with Grillo's message: a contempt for a duplicitous society whose moral righteousness is merely a mask, a performance but with this too: the Five Star Movement's "purity myth." The "purity myth" stems from creating Five Star Movement politicians like M5S deputy prime minister Luigi Di Maio as the embodiment or "the face" of uncorrupted virtue, antithetical to dirty Italian politics and collusions in it, able to usher in the "Movement from anti-politics to politics" (Minuz 2018). Grillo, too, fashions the purity of the movement by suggesting it chose Saint Francis as its patron saint in order to signify "politics without money and respect for animals and the environment" (Minuz 2018).[16]

To followers, Di Maio appears to be a troublingly perfect combination of "anti-politics webmaster" sporting "ties one wears to a small town wedding" and having an "old fashioned air about him even when he says 'I want a smart nation Italy'" (Minuz 2018). Di Maio has even been perceived as having been "programmed" as an anti-Renzi figure, a "projection of the Italian unconscious, an algorithm, a replicant, a populist android" (Minuz 2018). Indeed Di Maio emerged as a leader through voting procedures on Rousseau, so his candidacy was fundamentally a reflection of direct democratic but also algorithmic processes that might select for his seemingly perfect display of likability (D'Alimonte 2019). But Minuz is questioning his emergence as if he were himself "an algorithm," technologically fashioned from the preferences of the same body that later elected him leader. Di Maio is an algorithmic political leader, both insofar as algorithms are increasingly tethered to individual identity and from a political standpoint because he was created from more accurate prediction of individual choices and practices (van Dijck 2014). Big Data, we are told, ushers in "posthumanist ontologies" insofar as technical processes become deeply embedded in governance and society (Chandler 2015, 835).[17] Di Maio and the Five Star Movement's adoption of a digital platform represent a new posthumanist form of governance: a "populist android" leading an "algorithm party."

Algorithm Democracy

Media skepticism and supernatural discourse were surprisingly foundational in the creation of a party based on the "counterinformation" of the Internet; a high-tech algorithm platform, Rousseau; and a successful populist party (Loucaides 2019, 87). Rather than forefront the political ideology of the Five Star Movement, I have tried to retell its story by attending to the materiality of knowledge production. Grillo dismantled his followers' trust in media and government and channeled their faith into web-based communication and a never-seen-before digital platform and successfully installed a "new and updated version of authoritarian politics" (Caruso 2016, 587) and "a crowd-pleasing plebiscitarian system of governance easily exploitable by the few against the will and interest of many" (Navarria 2019, 209). How can we imagine the implications of this new techno-information knowledge-power posthumanist assemblage?

Besteman (2019) argues that roboprocesses result in "the proliferation of secrecy" because the circuits of power and capital become even more "opaque" (17). Besteman's insight helps us unravel why Grillo and the Five Star Movement's largely paranoid disposition toward institutions of power might be simultaneously what enabled their technological emergence.

Indeed, the very presumption of secrecy allowed for the new spin on Rousseau. Their deployment of an algorithmic political platform hinged on the fact that they did so transparently, and openly and enthusiastically invited party members to use and share information. But this message of honesty and inclusivity succeeds, I think, precisely because of their open criticism of surreptitious use of data and algorithms, as well as print and television media. Just as Berlusconi's gaffes and crude humor allowed him to consolidate power as a relatable figure, so, too, does Grillo's criticism of media and algorithms allow M5S to openly and successfully deploy the very same medium of information collection and analysis. In this sense the same political strategy of unhidden artifice or manifest disguise works well across the political spectrum; thus we must recognize its deeper epistemological roots, which we would miss were we to analyze it as merely an isolated political strategy.

Grillo masters the discourse of "algorithm transparency" (Besteman 2019, 22) even as the Five Star Movement has not been entirely forthcoming in how and what they mine or how they protect the data ("Illeciti sulla privacy" 2018). For example, they had initially used a faulty system in which the party member's phone number was used to verify electronic votes and the site was hacked in August 2017 ("Illeciti sulla privacy" 2018). The hacker had tauntingly teased, as "rOgue_O," "I'm still there" on Twitter even though the site was used a month later to elect the prime minister (Cuzzocrea 2017). Whether the data mining emerges from Google, Facebook, or even Rousseau, users do not necessarily have the resources or capital—social and otherwise—to confront the program makers about how and when their data are used. Consider the valuable data that might be collected from Rousseau users. Their patterns in searches and clicks, quantified time spent reading different issues, keywords searched or used might all be accumulated and analyzed to produce the most desirable party platform. It may be used to direct resources to the precise voter profiles and demographics based on age, region, or education level. And, just as Five Star deputy prime minister Di Maio has been accused of being himself an "algorithm," Rousseau may be used to choose candidates for office most aligned with constituent wishes and beliefs (Minuz 2018). While the algorithmic regime may offer opportunities for direct democracy—voting, law creation, event planning and protest—it may also prey on members' utopian gloss of Rousseau in order to manipulate their data and mobilize politically in covert ways.

Grillo masterfully plays both sides: he'll scare up Italian zombie stories and then turn around and have a fat zombie with a mohawk represent his political movement. He'll dance at Rousseau's feet and promise direct democracy with his digital platform and then spook his followers about the "reign of algorithms." Grillo's epistemic dualities reflect the two sides of information society: its deceptive deformities and its radical political potentials.

Chapter 4

The Trial against Disinformation

A little, light, long-drawn cloud . . . like a long very straight line.

—Aristotle, warning signs of an earthquake

His mood was solemn as he gazed on a map of his hometown, L'Aquila, Italy. Antonio was recounting the night of the 2009 earthquake: "I was lying in bed and thinking about the tremors of three possible scenarios and what I would do. First, where the load-bearing walls were, second, whether I would cross the room or get under the bed, and then, how I would cover my wife's body with mine." When the earthquake began only moments later, he only had seconds to shield his wife. He lay over her as chunks of the ceiling and wall fell on top of them, a piece of which tore so deeply into his leg that it exposed bone. This wound, he said, pointing at his shin, was noticed only after he'd spent hours trying to help others escape their buildings into the piazzas and secure medical assistance. He met my gaze: "These are the scenes that get under your skin."

Antonio and his wife were among the lucky survivors. The earthquake in the small Italian town of L'Aquila on April 6, 2009 was a devastating one: it was 5.8 Richter in magnitude, killed over three hundred people, "damaged 11,000 buildings and left 65,000 people homeless" (Iovino 2016; Pisa 2010).[1]

It took two years, from October 2012, for the earthquake to become known around the world, when the six scientists and one public official who had assessed risk were charged with involuntary manslaughter. The federal case alleged that these scientists were responsible for providing public reassurances that directly led to citizens' decision to stay home and led to their deaths. In November 2013, the six scientists of the Commission for Major Risks (CGR, *Commissione Grandi Rischi)*and one government official were found guilty, sentenced to six years in prison, and ordered to pay the equivalent of ten million dollars in damages (Ciccozzi 2016).

When the sentence was announced, crowds of people in the courtroom screamed "Shame!" (Caporale 2014). Newspapers soon reported that the scientists were charged for "failing to predict the earthquake" and employed imagery of Galileo in early stories about the trial (Hasian, Paliewicz, and Gehl 2014, 567). Scientific communities around the world were shocked and leapt to defend the accused in what became known as the "the trial against science" (*processo alla scienza*) (Mauri 2012, Salvadorini 2013).[2] The phrase "failing to predict" became "blackboxed" in global media, as was the assertion as made by one of the accused scientists Giulio Selvaggi: "An earthquake is unpredictable" (Kolbert 2015). Prediction became the salacious discourse that obscured and hid the more complex scientific knowledge on the earthquake. More than four thousand Italian and international scientists signed an open letter to Italy's then president of the Republic Giorgio Napolitano and called the accusations "unfounded" and "expressed his solidarity" with the scientists (Caporale 2014). Alan Leshner, the then editor of the journal *Science*, called the trial "unjust and disingenuous" (Cartlidge 2012, 184).[3] Claudio Eva called the sentencing "a very Italian and medieval decision" (quoted in Hasian et al. 2014, 572). Even after the appeal and acquittal of the six scientists, many argue that Italians still believe earthquakes are predictable "but [scientists] just don't want to admit it" (Kolbert 2015). In June 2010 the International Union of Geodesy and Geophysics made the following statement: "It is shocking and unacceptable to accuse and legally indict scientists and members of a governmental panel because they failed to make a prediction of an extreme natural event in a particular place" (Yeo 2014, 410). Invoking Italy as the country of Galileo, Michael Halpern (2012) of the Union of Concerned Scientists wrote a blog titled "Italian Scientists Jailed for Failing to Predict Earthquake." After the trial, the Accademia Nazionale dei Lincei published a letter attempting to correct this false but widespread view

Figure 4.1. Antonio gazes at a map of L'Aquila, June 2013. Photo by author.

of the case, which they called "misinformation": "The detractors distort the reality of facts."

The most dominant narrative of "failed prediction" relied on a several cultural assumptions with deep histories within Italy: century-long ideas that position Italians as naïve and backward, as overly Catholic, irrational, and pagan. These essentialized versions of Italian culture made it plausible that Italians expected scientists to be prophets who could perfectly predict earthquakes, which grossly distorted the court case and the legacy of the trial (Ciccozzi 2016). To assume the trial was based on irrationality obscures that what was truly audacious and entirely rational was taking seriously the enchantments of scientific, the tendency to view the "word of science" as holy and sacred or as tantamount to a supernatural force. It would also miss that predictions of 100 percent safety—the assurance that no earthquake would come—are indeed irrational and nonscientific claims. Finally, it would miss the cultural and political underpinnings of scientific absolutism (Ciccozzi 2016, 72). It was "science on trial" not because of a magical belief but because of the weirdness in viewing science as a modern form of magic.

Italy's "trial against science" was controversial in how it criminalized scientific prediction or, perhaps more accurately, scientists' endangering citizens' wellbeing and safety. What cultural beliefs about prediction, public safety, and scientific accuracy undergird this judicial process and conviction? Why did the trial emerge in Italy and not following natural disasters like Hurricane Katrina or Sandy, or the Tōhoku earthquake that led to the Fukushima disaster? In order to understand why "false" information was understood as deadly, this chapter aims to analyze how the initial event—the press conference and safety claims—and subsequent trial were informed and shaped by cultural perceptions and scientific understandings of disinformation, as well as earthquake prediction and risk assessment (Tipaldo 2015). Catastrophic disasters bear unique significance. As earthquake and art analyst Serenella Iovino frames it, "A world is undone, while another one enters the scene" (Iovino 2016, 85).

I argue that the context of widespread disinformation and socioeconomic precarity in Italy shapes why scientific claims about safety in the L'Aquila trial were viewed as criminally dangerous and led to the juridical progressive move to hold scientists accountable for disseminating misleading public information. The most significant context for understanding this trial is the culturally conditioned urgency to rescue and hold accountable scientific fact at a moment in which true facts appear on the brink of extinction. In other words, Italy's age of political enchantment, pseudo-truths, and media artifice plays a critical role in why scientists' "false reassurances" were criminalized. L'Aquila thus provides a way to think about how responsibility for natural disaster is legally managed and culturally perceived (Ciccozzi 2016; Pietrucci and Ceccarelli 2019). The government's emergency management in the immediate aftermath of the earthquake led to authoritarian forms of governance under a form of exceptional rule (Agamben 2005; Klein 2007; Guggenheim 2014). How might the epistemological context of disinformation and its reaction—holding speakers of misleading truths accountable—also shape why and how emergency governance takes hold? The trial in L'Aquila, though beyond the usual scholarly scope of disaster management, might also be framed as a kind of "exceptional" juridical intervention.

Collier and Lakoff (2015) argue that the notion that governments must prepare for societies that are "vulnerable to catastrophic disruption" is actually a quite recent phenomenon (20). Such government forms are sustained by different forms of knowledge, and the emergency governance or "vital systems security governance" is shaped by "system-vulnerability thinking, knowledge about interdependencies and vulnerabilities of vital systems" (23). Thus, to Collier and Lakoff, discursive production about potential harm in complex systems undergirds disaster management. While, like other postwar liberal democracies that they suggest shifted toward a recognition of "threats that were understood to outstrip the capacities of existing population security mechanism," the Italian state is also informed by this kind of risk-mediating knowledge, I instead highlight the role of disinformation in emergency governance (24). In the case of L'Aquila, we will find that the local government's anticipation of public unruliness and panic, a kind of predisaster emergency rule, put pressure on the scientists to issue false reassurances. The public panic grew in large part due to a slew of conspiracy theories about earthquake knowledge and prediction. In this sense, disinformation is an indirect though significant contribution to the installation of regimes of emergency governance and the trial against disinformation emerged because of both exceptional disaster governance and its emergent legal form.

Theorizing Scientific Prediction

The question of how temporal events, particularly human loss and death, are caused has long been the object of anthropological analysis with a fundamental truth that causal reasoning reveals underlying social understandings and epistemological frameworks for meaning-making (Parsons 1942; Jules-Rosette 1978; Nuckolls 1991; Guyer 2007). Cause-effect relationships, and the methodologies to determine these relationships, represent the core work of both science and magic and the distinctions therein. The notion of prophecy as "public discourse with a future-oriented social or political message" does not fully disentangle how foretelling in "non-ordinary" realms differs from scientific prediction (Leavitt 2000, 201). Among the kind of causality narratives is the category of present and future relationships, or foreknowledge. Parsons (1942) distinguished divination from science as a matter with the scope of people and events and the process by which it is made; moreover, failure in science is "due to faulty procedure" because something went awry in the divining procedure (339). Because the ritual of foreknowledge is at fault, Parsons outlines, in ways parallel to Evans-Prichard's (1937) conclusions on Azande oracles, that divination "can never be checked by reality" (339).

Man-made disasters—train accidents, flight crashes, and fires, for example—have often provoked charges of manslaughter (Alemanno and Lauta 2014, 143). However, the issue of criminal liability for natural, not man-made, disaster victims has grown greatly in part because the Act of God no longer exonerates human agents or institutional culpability (Alemanno and Lauta 2014, 142). Of course, the distinction between man-made and natural disasters has also become increasingly fraught. The judicial pattern, around the globe, has been toward an investigation of culpability: from floods in Saudi Arabia to tsunamis in Chile to mudslides in Russia (Alemanno and Lauta 2014, 143–144). In this sense, we must not view L'Aquila as an isolated incident or fluke, but rather as part of a much larger pattern in attributing blame to a variety of professionals and state institutions in the wake of natural disaster (Arcuri and Simoncini 2015; Benadusi 2016).

The European Court of Human Rights, in fact, requires a legal inquiry into accidents and disasters and serves to safeguard the European Convention on Human Rights Article 2 (the right to life) (Alemanno and Lauta 2014, 144). In 2008, the European Court of Human Rights ruled that "mismanagement" of a natural hazard might constitute a human rights violation from a case involving an "inadequate" warning system for a mudslide in the Russian town of Tyrnauz (*Budayeva and others v. Russia*) (Alemanno and Lauta 2014, 144). Moreover, the legal record in Europe has precedents for previous earthquake trials: an earthquake in Turkey led to allegations of state negligence (Alemanno and Lauta 2014, 144). Because courts are required to conduct judicial inquiry of disasters in order to ensure the right to life, it might, therefore, have been an infringement to halt or avoid such investigations in L'Aquila (Alemanno and Lauta 2014, 144).

Gary Alan Fine (2007) examines Chicago meteorologists and the outcome of a poor forecast in August 1990. Fine situates meteorologists as part of a daily and professional search for truth and reveals that "verification is an organizational practice. The adjudication of truth is a collective practice on which confidence depends" (195). His work sheds lights on L'Aquila as he examines cases in which meteorological forecasts become failures. A severe thunderstorm and tornado warning was issued for southern Kane County, Illinois, but an F5 tornado with three hundred fatalities devastated the northern part of the region (Fine 2007, 91). The error resulted in a lawsuit (*Bergquist v. US*), which was aimed at the National Weather Service and vacillated between agentic and structural, between individual choices and institutional culture (94). It was later dismissed because it fit within a "discretionary function exception" of National Weather Service liability (Klein and Pielke 2002, 1795). Fine suggests that the aftermath did not lead to "over-response" even though the disaster "haunts" Chicago (93); in fact, it became an emblematic case not about weather's predictability but rather "as an indicator of the unpredictable" (93).

In the United States, the Federal Tort Claims Act (FTCA) grants immunity to the government, which protects the federal government against claims based on either forecasting or failure to issue forecasts (Klein and Pielke 2002, 1791). The FTCA allowed for government immunity in the case of Kansas River Flooding (*National Mfg. co. v. U.S.*), against the levee board following the New Orleans flooding (*Spencer v. New Orleans Levee Board*), and against the National Weather Bureau for predictions regarding Hurricane Audrey (*Bartie v. U.S.*). In the event of Hurricane Audrey, the news broadcast "no need for alarm that night" shaped the decision of Whitney Bartie's family to evacuate and resulted in the loss of his wife and five children

(Klein and Pielke 2002, 1793). Though the court did not hold the bureau responsible because it could not conclude beyond a reasonable doubt that the warning shaped Bartie's decision, the court "cautioned that future warnings should use emphatic language concerning the urgency of evacuation" (Klein and Pielke 2002, 1793). In rare cases have damages been awarded to plaintiffs: a pilot was awarded $1.4 million because the National Weather Service had not corrected its inaccurate forecast of wind shear conditions that caused his airplane crash (*Springer v. U.S.*) (Klein and Pielke 2002, 1797). In 2010, Chilean officials were charged with manslaughter after failing to alert citizens to a tsunami, and Dutch officials were found guilty after ignoring climate change risks in 2015 (Benadusi and Revet 2016).

Prediction versus Prophecy

In a climate of fake news, and monetized "clickable" stories online, there are plenty of stories about unlikely or unusual practices for predicting earthquakes. According to the *Guardian*, the behavior of toads could have provided a necessary clue to the earthquake. Dr. Rachel Grant of the Open University had been studying toads seventy kilometers away and noticed their disappearance shortly before the quake. Toads, Dr. Grant commented, "are able to detect pre-seismic cues, such as the release of gases and charged particles, and use these as a form of early warning system" (Dollar 2010).

In *Scientists as Prophets: A Rhetorical Genealogy*, Lynda Walsh examines how scientists become positioned as experts capable of forecasting the future. For starters, she clarifies that she is not tracing the history of scientists as a whole but rather those who advise publics, a "scientific adviser" (ix). To Walsh, the "prophetic ethos manufactures political certainty in times of crisis" (Walsh 2013, ix). The notion of the "prophetic ethos" is the idea of inhabiting a sense of prediction and authority, which, furthermore, is often at the behest of politics or "manufactured" (2). Walsh, then, sees how the scientific predictive knowledge is shaped and conditioned by political forces and is itself a moral production based on unsteady and unpredictable conditions with a particular and especially politically desirable effect: stabilization. The prophetic ethos was first, before the seventeenth century, appointed to men of religion, yet first natural philosophers and then scientists came to inhabit this role (Walsh

2014, 3). To Walsh, foreseeing the future is only one aspect of prophetic work; instead, prophets instill and renew a society's core moral and political values of a particular society or "covenant values" (5).

While Max Weber has famously called this transformation from the religious to scientific leader the "disenchantment of the world," Walsh, informed by Latour and others, draws fewer distinct divisions between science and religion and modern and premodern (7). Walsh argues, "The way we treat *science advisers* is evidence that our political world has never truly been disenchanted" (7, italics in original). The treatment of science advisers is, too, unpredictable and "oscillates between worship and witch-hunt" (7). The media play a crucial role in creating the prophetic ethos and, occasionally, tipping it into cultlike worship of science, transforming the advisers into Oracles (149). She notes that climate change scientists are among many lobbyists and experts waging a "prophetic battle" about future climates and who becomes authorized to predict and proclaim (163).

The Risk Commission's Predictions

We must now go back a bit further to understand L'Aquila as a historically seismic zone, with recorded earthquakes since 1349 (Cartlidge 2012, 185). In 1985 and 1995, they had what have been called "swarms," or small tremors, which occurred over a month's time but without being followed by any major earthquake. In early 2009, L'Aquila again experienced swarms, and it was during those weeks that the Risk Commission spoke to citizens. This became criminally actionable when some of those citizens failed to properly evacuate their homes.[4] The seven defendants were held responsible only for those deaths, numbering thirty-one, of people who were proven to have reentered their homes based on assurances from the commission (Cirillo 2013).[5] The biggest sources of evidence came from sound bites in the newspapers and in their televised press conference: members of the Risk Commission were quoted as saying that swarms "were *never* a precursor of large seismic events" (Cartlidge 2012, 186; Pietrucci and Ceccarelli 2019, 110). The vice president of the commission said swarms "didn't forecast anything" (Cartlidge 2012, 186). Now, the scientists were drawing from evidence that shows that swarms have been followed by major shocks only 2 percent of the time and have not been reliably used to predict major earthquakes (Kolbert 2015).

Another member of the group went on television and said, "The scientific community continues to confirm to me that, in fact, it is a favorable situation." An interviewer asked him if people could relax at home with a glass of wine, and he responded, "Absolutely" and recommended a good Montepulciano (Cartlidge 2012, 185).[6] Perhaps most bizarrely, one of the scientists on the commission, the head seismologist, had actually authored a paper that said an earthquake in the region before 2015 was certain to happen (Cartlidge 2012, 187). Yet he still seemed secure in foretelling L'Aquila's safe future. Moreover, there was a great cultural work in establishing the members as knowledgeable, trustworthy, and authoritative speakers of truth. For example, they were called "the highest scientific authorities of the seismic sector" who are "able to provide the most current and reliable picture of what is happening" (Barberi et al. 2009, par. 3). Barberi himself also described the meeting as "to make an objective assessment of the seismic events taking place in relation to what can be predicted" (Barberi et al. 2009, par. 8).

In 2012, *La Repubblica* released a conversation between Bertolaso, the head of the Civil Protection Department (Dipartimento della Protezione Civile), and a local official in which Bertolaso reveals that the press conference was strategically designed to soothe the public, which was especially roused because of Giuliani's prediction and visibility: "I will send them there mostly as a media move. They are the best experts in Italy, and they will say that it is better to have a hundred shocks at 4 Richter than silence, because a hundred shocks release energy so that there will never be the big one" (Yeo 2014, 408). In his sentence, Judge Billi referred to the press conference as a "media operation," in an indirect reference to Bertolaso's leaked comments here (Yeo 2014, 409). Therefore, this event raised skepticism because Bertolaso appeared to devise and design the press conference as a way to purposefully subdue the anxious population. Moreover, he directed the citizens to believe in the disinformation regarding a "release" of energy that made shocks appear to predict no further earthquake.

Vincenzo Vittorini, a local surgeon and trial witness, said, "It was as if we were *anesthetized* like someone had removed our primitive fear of an earthquake. After that damned [press conference], they instilled in us the idea that something terrible couldn't happen" (Cartlidge 2012, 186). The metaphor of anesthetization recalls an act of temporary numbing, and, like suspension, it implies a temporal altered reality and a psychic dissonance.[7] Žižek (2006) suggests that "the illusory perception of scientific discourse is that it is a discourse of pure description of facticity" (164). Part of what interests me here is what cultural structures and conditions help achieve this anesthetizing, as only certain knowledge can numb, and only certain kinds of actors and their racialized, classed, and gendered bodies get believed without doubt. In fact, the numbing goes further than just belief in scientific fact; it implies a process of making that authority deceptively powerful. This illusion that science is "pure fact" is precisely what makes it so authoritative. If we follow the enchantment theories, then we might conclude that the L'Aquilans were conditioned to believe in staying home because of their priests or their superstitious belief. But the decision was made because they listened to and believed in scientists, a social fact that rested on a particular configuration of how scientists become reliable truth-tellers. In fact, there appears to be an excessive faith in or even an exploitation of this position of scientists as believable that had nothing to do with Italians' belief in magic or the occult. Put simply, it is not just about enchantment but a crisis of enchanted science which sustains an essentialized discourse of Italian gullibility from being too Catholic or too pagan, too unmodern, and now, too believing or disbelieving in science. Indeed, Claudio Eva, one of the convicted scientists who hailed from the University of Genoa, called the court decision "very Italian and medieval" (Kolbert 2015).

The court's criminalization of "false reassurances" emerges in a context in which false information is very rarely seen as having deadly consequences. In other words, I am arguing that the trial is an attempt to hold people accountable for disinformation in the age of Berlusconian and media misinformation. The trial was shaped by both socioeconomic precarity and epistemological precarity in which the truth of science has higher stakes. What was truly audacious and entirely rational was the courts' taking disinformation as well as the tendency to view the "word of science" as holy and sacred or as tantamount to a dangerous force that acts on human behavior. Science was indeed "on trial" not because of a magical credulous belief but because the court positioned science as a body of knowledge that gets mystified and shapes human action in irrational ways. In a time when there are seemingly few repercussions for disseminating lies on Italian television and in the piazzas, the courts actually held individuals accountable for their pseudo-truths.

Now, the trial witness—the metaphorically anesthetized surgeon—trusted the scientists' soothing words. He and his family stayed at home and he lost his wife and daughter (Cartlidge 2012, 186). To what extent was this because he understood a kind of mystical authority to science as "pure fact"? Would greater skepticism regarding the scientists' assurance of safety have saved his loved ones? Another trial witness likened the words of the commission to words "from heaven" (Cartlidge 2012, 187). Antonio, who recounted his survival to me, blamed the scientists: "The scientists needed to do their jobs, interpret scientific information and pass it from scientific institutions. . . . This is seismic territory: a catastrophe might happen after a seismic tremor [*sciame*], so you cannot say that nothing serious will happen."

Antonio's reaction shows his skepticism about any avowal of certainty and safety from scientists, questioning the very premise of this kind of prediction. In 2010, the prosecution letters were sent to the seven members of the Risk Commission, in May 2011 was the formal indictment, and the trial began in September of that year (Yeo 2014, 408). The international media coverage of the trail fixated on the "failure to predict" notion positioning Italians as gullible and as a population that saw scientists as prophets. The wording of the criminal sentence attempts to correct these misunderstandings. Judge Billi's sentence declares, "The work of the defendants (*imputati*), was certainly not to predict the earthquake ("Sisma" 2013)" but rather "[their] poor risk assessment and presentation of incomplete, falsely reassuring findings to the public" (Cartlidge 2012, 184; Case no. 380 22.10.2012, *Victims of the Earthquake versus Barberi et al* (members of the National Commission for the Prediction and Prevention of Major Risks), Tribunal of L'Aquila.[8] Citing the "incomplete, imprecise, and contradictory information on the nature, causes, and future developments on the dangers of seismic activity in question," the judge reasoned that predicting a lack of event with unreasonable certainty criminally threatened public safety (Alemanno and Lauta 2014, 139). Moreover, the scientific community, specifically, the International Commission on Earthquake Forecasting for Civil Protection (ICEFCP), disagrees with the Risk Commission's claim that seismic activity does not predict earthquakes; in fact, they indicate that tremors marginally and potentially significantly indicate a stronger quake (Yeo 2014, 409). During the press conference, Barberi had said, "There is no reason why we can say that a sequence of aftershocks of low magnitude can be considered a precursor of a strong event" (Yeo 2014, 409).

The judge punished the scientists for their unscientific knowledge, their seemingly Godlike security in suggesting *nothing* would occur. In a certain sense, this sentence values transparency and rational logics: making poor scientific misinformation in the media criminally actionable. Indeed, my work has found evidence that the Italian courts, by showing neoliberal and Catholic values, might be uniquely progressive. In *Labor Disorders*, I examine how courts began awarding a new category of damage for workers who were psychologically harassed and isolated: existential damages (*danni esistenziali*), which was more metaphysical than an existing category of moral damage (*danni morali*). Here, the courts seemed to recognize a kind of ineffable and intangible injury of Italy's capitalist labor regime, so perhaps, I wonder, might they also be one of the first judicial systems to take media's disinformation to task?

During the trial, a local anthropologist, Antonello Ciccozzi, testified as a witness regarding the causal relationship between authoritative knowledge and people's actions, or, put differently, particular utterances and the victims' decisions on the evening of the quake. He (2016) writes that role in the court was two-fold: "My task was to demonstrate that some (some, not all) L'Aquila citizens would have survived had they not been persuaded that no destructive earthquake was to occur that night" (70). He testified that the L'Aquila citizens' understanding of the scientific discourse was "a form of supreme and unchallengeable truth in Western culture" (70).

Ciccozzi's testimony included the following: "In Western society scientific authority is perceived by the masses as the highest expression of authority, which . . . condition[s] collective action" (Case no. 380 22.10.2012, *Victims of the Earthquake versus Barberi et al* (members of the National Commission for the Prediction and Prevention of Major Risks), Tribunal of L'Aquila Sentenza N. 253/2010/380/2012). Ciccozzi was trying to make the case that scientific information was the indisputable causation for the thirty-one victims' decisions to forgo safer accommodations. Ciccozzi has since imagined the trial as a kind of "epistemological ritual that stage[s] and spectacularize[s] specific manifestations of authority" (68). Ciccozzi illuminates how the court is shaped by "the legacy of positivist approaches" and "the myth of absolute objectifiability" and thus the trial becomes a ritualized opportunity to avow a singular belief in absolute scientific truth (68).

Yet here we must pause to read into this unfamiliar event. Why did the court turn to an anthropologist to verify belief in science? It happened as if the belief in science were as exotic or strange as belief in witches, prophets, or gods. In fact, science is regularly coded as knowledge, not belief; the difference is that "knowledge" has come to mean information we confront with certainty and security, whereas "belief" has come to imply doubt. To say one "believes" in God once predominately meant a pledge—as the word "belief" has roots in the words "love" (Good 1994). Yet to say one believes in God today implies the possibility of doubt. Indeed, the presence of the witness also implies that other belief systems were mitigating how actors approached "scientific thought," or, put differently, that Cicozzi indirectly positioned scientists alongside other masters of belief: priests, fortune tellers, and astrologists. Doing so reminds us that scientists' authority is not, in fact, absolute and eternal but rather dependent on particular cultural, historical, and material circumstances.

In June 2014, I interviewed a local physicist and fellow L'Aquilan, Marco, who joked and said Cicozzi was wrong. Instead, Marco argued that he, like most Italians, never really listened to scientists and mistrusted pretty much everyone. I must say Marco's theory also seems somewhat reasonable: the prevailing deep cynicism means that Italians have a significant doubt toward anything packaged as authoritative knowledge. At the same time, Ciccozzi's recognition of scientific authority's powerful believability is equally compelling. When I spoke to Ciccozzi himself in L'Aquila in June 2013, he told me that he had been facing death threats since he testified. It appears to me that Ciccozzi was threatened because he took seriously that scientific truth—regardless of its veracity—must be recognized as a sometimes irrational cause of human action; he dared view scientists as criminals. Moreover, he treated science as a discourse of belief, one among many domains of knowledge that shape human behavior and produce direct causal actions. Finally, the danger in his stance, together with the court, was that he insisted on scientists' individual responsibility for disinformation. This is audacious considering this was accountability for misinformation in the context of a wild and irresponsible culture of pseudoscience and pseudotruths. It is precisely this issue that tells me that we must study the ways in which misinformation and spectacle shape the veracity in science, politics, and news media. In mid-November 2014, an appeal trial overturned the sentence and absolved the scientists, a verdict met with anger from local citizens and relief from the scientific community. It upheld the conviction of Bernardo De Bernardinis, the DCP official who reported the scientists' conclusions at the press conference, which thus "divides risk analysis from risk communication" (Pietrucci and Ceccarelli 2019, 97). The court's failure, however, is less surprising when we consider that public disinformation has become the cultural and historical norm in Italy.

The threats to Ciccozzi might also have to do with how the trial impeded postdisaster governance. In other words, the state might have an interest in maintaining public order, both before and after disasters, and making each public news event criminally accountable would transgress the order of disaster capitalism and a "state of exception" (Agamben 2005). Agamben (2005) examines how such forms of exceptional governance have become increasingly common and draw from an extended range of trigger events including humanitarian, ecological, and natural disasters, as well as domestic strife and international threats (Dillon 2007).

Italy has a 1992 civil law that allows for martial law and for public funding to be funneled into private hands provided that situations have "emergency" status or, in the case of L'Aquila, are considered a "big event" (*grande evento*).[9] Functioning in ways similar to "states of emergency," "big events" suspend civil law yet have typically been used for events like a pope's visit, not disaster. But "big events" require less bureaucracy and urgency than "states of emergency," which makes them more dangerous because they can easily suspend democratic governance (Calandra 2018). The first step was Prime Minster Berlusconi's declaration of a national state of emergency. Following that declaration, the health care commissioner, who was a member of the National Healthcare Service (*Servizio Sanitario Nazionale*), suspended the Budgetary Recovery Plan such that expenses did not have to follow the preordained limit (Sargiacomo 2015, 73). Other exceptional practices were allowed, which included allowing hospitals and pharmacies to distribute necessary prescription drugs without official authorization and to give over-the-counter drugs at no cost (Sargiacomo 2015, 74).

Once L'Aquila was deemed a "big event," civil law was suspended, and this "exceptional" moment gives way to authoritarian and militarized governance.[10] Certain areas of the city, including the historical center, immediately became emergency zones where no people were allowed (Calandra 2018). Survivors were accommodated either in new residential complexes or temporary housing units on the outskirts of the city.[11] In terms of rebuilding L'Aquila, this legal shift enabled rebuilding in areas previously considered illegal and suspended building codes and law. L'Aquila raises questions of how global disasters become a means for suspending democratic order, militarization, and more broadly, shaping forms of governance.

ANTONIO'S REFLECTIONS ON THE TRIAL

In June 2013, Antonio and I spoke at length about the trial and its legacy.

Noelle: Has there been a culture of fear around earthquakes and preparing for them? I've been told, and I'd be curious what you thought, that there was a moment before the 2009 earthquake when people weren't afraid because they were reassured, which was strange, because there had been a thousand-year history of fear and reaction.

Antonio: Yes, this is what was also reported in international news, the fact that it was written that the scientists were punished because they said they couldn't predict earthquakes. It's a scientific reality that you can't predict earthquakes and that is correct. The problem is that just as you can't predict earthquakes, you can't predict that there won't be earthquakes. They said people should just relax and they said that it was better because the earth was emitting energy, yet that wasn't the case.

Noelle: But do you think people listened to these scientific authorities, and stayed home?

Antonio: I didn't stay home, I didn't leave, I didn't minimally think that there would be an earthquake, beyond any shadow of a doubt.

Noelle: But I mean, from your experience, your sensibility, or from their reassurances?

Antonio: I have to say, maybe it was just me because I had never had the experience of a big earthquake. But I never thought that something like that would happen, I just never even imagined it. But my, let's say, my superficiality, plus the fact that they said to just relax, I didn't feel like . . . it just didn't make sense for them to say relax. If they had to there were places where people could meet or unite—this would have let people understand that there might be something of a certain intensity. In terms of prevention, that night I didn't sleep at home, but it's difficult because maybe it doesn't happen that night, and I don't think people would stay outside of their

> homes for four or five days. . . . The problem was that nobody was out for a trial against science. As a scientist myself, I don't think that they, well, they needed to do their job, basically. Everyone interprets reality as you want, but the interpretation of scientists has a value, a scientific value in and of itself. You can interpret these scientific data, I mean, you can pass them to an organizational and administrative sphere, like, politics, but I don't think you can expect approval from other scientists. . . . The scientific reality that emerges from data is that L'Aquila is a seismic zone, there have been earthquakes in the past, there can be a catastrophic seismic event after a seismic swarm. You can't say to just relax and the earth is emitting energy; you also can't say that the fact that this was happening signifies nothing. Yes, not every swarm happens before an earthquake, but you can't say it means nothing.

Magic and Science on Trial

In responding to the vital mystery of how L'Aquila is Italy-specific and how the justice system and law are culturally understood, I have searched for Italian precedents, real or literary, for putting science or belief on trial. A 1917 play by famous Italian playwright Luigi Pirandello, *The License* (*La patente*), tells the story of Rosario Chiarchiaro, a *jettatore*. The term *jettatore* loosely translates as a "master of misfortune" or, more literally, a man who distributes *jetta* or bad luck to people (de Ceglia 2011, 76, 80; de Martino 2004, 153).[12] In the play, Chiarchiaro seeks to transform his vocation into an actual form of employment and petitions the courts for the play's namesake: an official state license to practice (Simioni 1970). Actually, he wants the license in order to collect a fee from citizens for not harming them; naturally, he insists that such a tax is not extortion but rather legal compensation he is rightfully due.[13]

Italian ethnologist Ernesto de Martino was also fascinated by the *jettatore* in his studies of magic and belief. He argued this figure called into question the rational actor for whom actions move with "regular predictability" (de Martino 2004, 154). Part of the *jettatore*'s fictive trial, like that in L'Aquila, was predicated on what counts as rational forms of foreknowledge and public expectations for prediction. It has long been a truism in the study of magic and science that causality is precisely the turf around which magic gets differentiated from science: a seed grows into a plant, yet the causality of how those events are related often divides magic, religion, and science. In both fictive and real worlds, the court, as the symbol of high civilization and rationality, must clarify who is officially licensed to foresee the future, how we distinguish prediction from prophecy, and why certain misfortunes are unevenly distributed. In other words, the trial in L'Aquila was also trying to make sense of why thirty-one people died. Why them and not thirty-one other people? The answer might have once been seen as an act of God, as a sign of divine punishment (Alemanno and Lauta 2014, 138), or maybe because the *jettatore* had cast bad luck their way. But today the word of science is positioned as the site of blame. Therefore, the court is creating a narrative about how to understand the randomness of death and human suffering.

In his work, de Martino also emphasized the embodied experience of belief. For de Martino, the *jettatore* represented a paradox between thought and action, what he called "It's not true but I believe it" (de Martino 2004, 154); that which simultaneously enacts belief and unbelief, what he names "a no-man's-land between the true and the false" (de Ceglia 2011, 88).[14] De Martino's "no man's land" resonates deeply with L'Aquila and seems to encapsulate the institutionalized blurring between the made and the made-up in Berlusconian Italy.

Earlier I struggled between Marco and Ciccozzi's version of the story: did L'Aquilans trust nobody, or did they blindly trust scientists? We can get insight by considering De Martino's "It's not true but I believe it." In L'Aquila, citizens enacted a sense of "Predicting 100 percent safety is not true but I believe it."[15] Still, the notion that Italians would have believed in scientists' irrational and prophetic power seemed plausible to the international community. Yet what was actually quite radical was that the court was examining skepticism and doubt in science.[16] It was not that Italians believed science could predict an earthquake; it was that Italians dared to hold scientists legally accountable for disseminating bad science. Here, the science of risk became a kind of "mythologized reality" that reified scientific knowledge.

Vampire Earthquakes and Suspending Law

The L'Aquila earthquake produced a variety of highly visible counternarratives, yet these also seemed to work on a basic epistemological premise that "the truth" was submerged. Sabina Guzzanti, political satirist, Berlusconi impersonator, and film director, released a documentary about L'Aquila in her 2010 documentary, *Draquila: Italia che trema* (*Draquila: Italy Trembles*), playing on Dracula.[17] The documentary's supernatural imagery should not go unnoticed: the reinvigoration of the supernatural into political discourse is a manifestation of reenchantment, and my research has tracked this noticeable pattern of populating political life with zombies, ghosts, and vampires.

Guzzanti's film begins with the context of then prime minister Berlusconi's "shitty day": he was under criminal investigation for corruption, mafia connections, and underage prostitution and had nose-diving approval ratings. In the next scene, Guzzanti visits L'Aquila dressed as Berlusconi and says, "This earthquake has been a grand success! No other earthquake has given us the discounts that this one has given us!" The film argues Berlusconi used L'Aquila as a way to amend his damaged reputation and corruptly rebuild the city. He had already received international attention for quickly moving Aquilans from tent cities known as "tentopoli" to antiseismic apartments on the outskirts of town which were called "new town" and later dubbed "Berlusconi's homes" (*le case di Berlusconi*) (Dollar 2010). The film portrays L'Aquila as a tragedy of disaster management and the corrupt coopting of science by politicians. In addition to an elaborate housing development, Berlusconi managed to host the G8 summit in L'Aquila and pour millions—over 180 million euros—of public funds into mostly private hands (Cerasoli 2010, 41).[18] Complete with rapid road and hotel construction for visitors and a slogan draped over one of the camp cities—"Yes we camp"—it has been called a "brilliant operation of political marketing," which indeed fits with the culture of Berlusconian political theater (Cerasoli 2010, 42). In this sense, Guzzanti's story is not about the scandal of believing in scientific warnings but rather the corruption of scientists as traitors—self-interested calculators who assured the population for profit or, at the very least, did so as a political favor. It is a far more cynical rendering than scientific fallibility and the naïve suspension of skepticism.

The documentary also focuses on the political manipulation underlying disaster management. A 1992 civil law (Article 5 of Law 225) allowed for

Figure 4.2. A street in the abandoned city center of L'Aquila, June 2013. Photo by author.

martial law and for public funding to be funneled into private hands provided that situations have "emergency" status or, in the case of L'Aquila, are considered a "big event" (*grande evento*).[19] Functioning in ways similar to "states of emergency," "big events" suspend civil law yet have typically been used for events like a pope's visit, not disaster. Once L'Aquila was deemed a "big event," civil law was suspended in the "exceptional" moment and gave way to authoritarian and militarized regimes of governance.[20] From the first night after the earthquake, the city was under military rule and the "tentopoli"—the camps constructed to temporarily house Aquilans after the quake—were under great public scrutiny (Cerasoli 2010, 37). The city's accounting practices in terms of health care provision were also modified as a result of the emergency intervention; like forms of exceptional governance, these practices have been framed as "extraordinary accounting" in order to call attention to the deployment of calculative technologies sustained under emergency events (Sargiacomo 2015, 68). Sargiacomo (2015) elaborates on how exceptional governance thus relies on forms of quantification as the government relied on creating new classifications and practices as well as authorizing experts and advisories to define costs and funding needs. For example, a new system was designed in order to distribute medications along with an identification code which would be carefully surveilled according to aid regulations (74). The territory was redefined to classify individuals eligible for benefits and housing as opposed to normal citizens.

In L'Aquila, Civil Protection Services made life in the tents quite rigid and increasingly hypercontrolled. Residents were not allowed to gather, to circulate leaflets, or even to move freely to other camps (Cerasoli 2010, 39). Citizens also became suspicious about "new town" as an *arriére pensée*, an ulterior motive to exchange to avoid rebuilding the historic town center for the housing development (Cerasoli 2010, 40). The housing developments are designed to be semipermanent and not, at least in theory, replace long-term plans to rebuild L'Aquila. In June 2009, over a thousand citizens went to Rome to demand the full reconstruction of the city (Cerasoli 2010). When the new housing became available, Aquilans were "divided between the Cassandras who were prophesizing the refusal to reconstruct the city and the approvers who, with the motto 'Thanks Silvio', welcome having averted the threat of [living in] containers, embark on a giant public contract, marked, in the name of emergency, by a drastic simplification of law" (Cerasoli 2010, 41).

At one point during the documentary, Guzzanti interviews a man about the ongoing national scandals surrounding Berlusconi. He replied, "I don't believe them; they're all lies." The interview here implies a level of popular incredulity toward political corruption. Not believing the copious amount of evidence presented against Berlusconi in his national trials seemed to become magically displaced onto not believing that Berlusconi or other politicians would orchestrate rebuilding the city merely for personal and financial gain. The film ends with not believing. Guzzanti interviews an older gentleman who says, "I've spoken to lots of people who lived under dictatorships and they tell me: the downfall of some people came when, after a few years, they keep repeating, 'Now it's gonna fall. This can't last.' This is big delusion: if it is empty and fragile it can't last. It's not true! It lasts [*dura*]." At this word, "lasts," the film ends abruptly: the screen eclipses to black.

Perhaps what L'Aquilan citizens suspended when they listened to those reassurances, then, was the truth of authoritarian power, and in the process, they become subjects of the millennial emergency state or "limbo agents" of Italy's truth society. Like a ghost that one believes in so the play makes sense, the ghost here is democratic governance and objective truth. It appears far-fetched, but the rest of the narrative makes sense only if the ghost exists. But the ghost, like liberal ideals of civil governance and transparent truth, hides what may be even harder to believe. Yet here they are slippages of consciousness or psychic hauntings: What L'Aquilans cannot believe in but sustains the narrative is the corporatization of human life, the co-opting of science by private interest, the undermining of human life by forms of paternalistic governance. The suspension of disbelief uncovers "the unexamined assumptions which are as much the product of demythologised realities as they may be of mythologised ones" (Kapferer 2007, 342). Here, the heroics of science and democratic governance become the "mythologized realities" with high stakes.

In return for his help enacting this law, Berlusconi nominated L'Aquila's head of Civil Protection, Guido Bertolaso, as national minister of Civil Protective Services—the very part of government that tips into action after a "big event." This same Bertolaso was arrested for manslaughter for his role in convening the commission, corruption, unaccounted spending, and even prostitution. The documentary charts the phone-tapping scandal, including one call in which Bertolaso said he planned to stage the Risk Commission's press conference as a "media operation" and explicitly asserting his desire "to reassure the public" (Cartlidge 2012, 187).[21]

Aftermath and Scientists' Reframing

After the appeal, the scientists have narrated and reframed the trial, once again, and exposed the manipulation of scientific discourse by Civil Protection, as a mechanism of population control. Enzo Boschi, a seismologist for the National Institute of Geophysics and Volcanology, said, "We scientists were used. What happened in L'Aquila on March 31, 2009 was a political meeting, as the lawyer Franco Coppi defined it. It was the choice of Civil Protection Services that wanted to respond to the alarm of imminent earthquake created by that technician, Giampaolo Giuliani. . . . I never would have done it. I wouldn't have given reassurances. . . . The mayor of L'Aquila, Massimo Cialente, was there and he witnessed that he was so impressed by my declarations of seismic risk during the meeting that he decided to close some schools and ask for a state of emergency" (Caporale 2014; "Sentenza L'Aquila/Giuliani" 2012). To his interviewer's reminder that he had called a larger quake "improbable," Boschi retorted, "Certainly, I explained it was improbable but not that you could exclude the possibility. The language of science is different from the language of media and communications in that it must be 'conveyed' [*veicolata*] to the population. It is the duty [*compito*] of Civil Protection to determine in which cases what is necessary to know. This is the law." The trial set a precedent to disentangle science from public media, holding only the media accountable (Pietrucci and Ceccarelli 2019, 98). Indeed, we may see that scientific disinformation—knowledge that passes as authoritatively scientific yet without necessarily passing the rigor and method of scientific knowledge production—becomes an increasingly important feature of disaster prevention and management.

Continual Contact with Reality

Let us make one final return to L'Aquila. In the light of late afternoon, Antonio gazed on a map of his city, his fingers brushed its surface longingly, and he said, "It was like we started over from zero; it's not how you remember, it's a precise cut, a sensation of fleetingness. The thing is, you feel like it won't be like [your] prior experience, because you live a kind of before and after, whereas before it was super-lived [*stravissuti*] and now it's just death." Though I'd only seen L'Aquila wrapped in wooden boards and canvas tarps, I was trying to understand. Antonio continued, saying, "The strange thing about earthquakes is that I wanted to see the destruction. Despite everything, I had a need, an extreme need, to see the fallen homes, look online for it, whatever—but to be in continual contact with this reality."

I sympathized with Antonio's "need" to be in "continual contact with reality," but I have also reimagined this need for "reality" as not only one survivor's grief but a complex feeling produced by ecological, social, and epistemological flux. His need is not only one individual's psychological coping but a need conditioned by a particular structural and material "reality" in which truths are regularly artificial and manipulated. Antonio's need for reality is like one visible tremor made by massive tectonic motion.

Chapter 5

Scientific Anesthetization in the Anthropocene

An atmosphere suspended by silent dust dominates everything, absence, emptiness. What is not destroyed, upon closer inspection, is totally dangerous. But penetrating the labyrinth of a place devastated by an earthquake is a feeling that is difficult to describe using words or even images: you can't fully convey the sense of total destruction, the fear of the apocalyptic inadequateness of human creations that rises from the surfaces of a place marred by nature.

—Antonio Ciccozzi

Giacomo's Drive toward the Middle of the World

It seemed like the road kept circling deeper into the mountain, funneling and zigzagging, one long dark corridor after another. The long tubular tunnels reminded me of other trips around Italy, with their long mountain tunnels. In those middle-mountain passageways a circle of light grows larger as you speed toward it. On some occasions, yes, that light disappears for a few seconds, maybe even a minute in some of the longer corridors. But it returns soon enough. This time, the car's steady movement coincided with no circular horizon, no break in the shadow and darkness. It seemed as if with each new corridor it was getting even darker and the space above us felt denser. How different it seemed for Giacomo. He drove with the ease of habit, the curves and darkness like roundabouts and traffic lights on one's commute. I wondered how a daily pilgrimage into such earthly depths might change how he saw the world, the universe. After all, the purpose of establishing a laboratory in the mountain was to approximate "cosmic silence" and create optimal conditions to study space, especially dark matter. What an odd and beautiful mirroring: burrowing deep into the earth to understand faraway darkness. Giacomo, a survivor of Italy's devastating 2009 earthquake in Aquila, was taking me on a tour of the enormous particle and nuclear physics labs located underneath 1,400 meters (4,600 feet) of rock, at the Gran Sasso National Laboratory, one of the world's largest underground research facilities.

When we arrived, he gave me a construction hat to put on and showed me the long tubes of backup lighting that ran along the hallways. The main halls were enormous, over sixty feet wide and tall and over three hundred feet long. Occupying the space were enormous cylindrical and rectangular machines, one of which holds the Oscillation Project with Emulsion-tRacking Apparatus (OPERA) instrument that detected subatomic particles, tau and muon neutrino oscillations.[1]To my untrained eye it was a large rectangular object with black and silver rails, with red and white panels striping down one side. The Imaging Cosmic and Rare Underground Signals (ICARUS) physics experiment also detects neutrinos and appears as a giant silvery white rectangular box. The towering space-sensing structures stood in stark contrast to the fluid curves that brought us there.

Disaster and Causality

Standing alongside ICARUS, I began thinking about how scientists and everyday people connect events, attribute causality between one event and another, and assign responsibility to something or someone. For earthquakes, scholars date the shift between the Enlightenment paradigm of blaming God toward a more secular understanding of blame to Rousseau and Voltaire's lived vulnerability from the Lisbon earthquake of 1755 (Alemanno and Lauta 2014, 138). Jean-Jacques Rousseau wrote, "I do not see how one can search for the source of moral evil anywhere but in man. . . . Moreover . . . the majority of our physical misfortunes are also our work. . . . It was hardly nature that there brought together twenty-thousand houses of six or seven stories" (Alemanno and Lauta 2014, 140). Rousseau was, of course, not finding causal nature in the earthquake itself but in the human design of the city that resulted in high fatalities. Rather than blame God for "moral evil," he attributed the blame to the manmade urban design. Still, it was a sign of secularizing the causality of disaster, as the human loss of life was not attributed to their moral failures or evil.

Today disaster analysts use the notion of vulnerability theory as a way to position the underlying social causes of loss and trauma and make sense of disaster: "In this epistemological regime, the interior of the disaster is turned upside down; as the understanding of what a disaster is, takes departure from the social itself" (Alemanno and Lauta 2014, 140). Judith Shklar deploys the distinction between misfortune and injustice in order to make sense of how blame is theorized following a disaster. Under a pre-Enlightenment notion of disaster, misfortunes were seen as being caused by "unaccountable external causes," whereas the paradigm attributing disaster to a "social system's weaknesses" can be recognized as an injustice with "blameable" subjects (Alemanno and Lauta 2014, 141–142).

In terms of L'Aquila, communication scholars Marouf Hasian Jr., Nicholas Paliewicz, and Robert Gehl (2014) suggest that the members of the Risk Commission had trouble not in the scientific understanding of earthquake risk and prediction but rather in communicating this complexity to the public: "They should have been clarifying their positions on aleatory and epistemic uncertainty in seismological contexts" (561). The discourse that the trial was somehow antiscience thus "downplayed the contingencies and partialities of scientific knowledge" (563). Part of the issue of "nature hazard forecasting," they suggest, pivots on the issue that no audience has a uniform notion of risk assessment or who is authorized to know and forecast (564). Part of this "epistemic uncertainty," I would suggest, is also due to the undermining of nature forecasting because of how man-made disaster and climate change, the notion of the Anthropocene, have interrupted the sense of what constitutes a natural disaster, culpability, and scientific ethos. As Hasian Jr et al (2014) rightly suggest, "uncertainty is often itself an ideological, rhetorical, and sociotechnical achievement." The sociocultural and rhetorical understanding of Anthropocene has directly and indirectly rendered cultural categories like "natural disaster" and "prediction" more fraught and thus destabilized the causality of disaster (565), and conspiracy theories represent a new understanding of human responsibility in environmental disaster and malady.

The Anthropocene

The Anthropocene designates a period when human activity propels ecological change and global climate more than other earthly forces and matters, and the historiographical period would come after the Holocene, which covers the past ten thousand years (Kohn 2014; Bauer and Bhan 2016; Sayre 2010). The term "Anthropocene" was coined by biologist Eugen Stoermer and then popularized by Nobel Prize winner and chemist Paul Crutzen. Scholars debate whether the beginning of the Anthropocene should be in the eighteenth-century Industrial Revolution, particularly when the steam engine was developed, or in the twentieth century at nuclear detonation (Bauer and Bhan 2016, 62). For some critics, the notion of period troubles the boundaries between human and nature, or "the nature-society and the natural-anthropogenic binaries" (Bauer and Bhan 2016, 62).[2] Lesley Head (2016) suggests: "The evidence of the Anthropocene requires us to rebuild its own conceptual scaffolding in order to imagine and enact the world differently" (15).

It is precisely the kind of imaginative and conceptual work that this way of thinking about the planet that might shape the Aquilan case specifically and understandings of earthquakes and environmental risk more broadly. Moreover, the trial allows for new ways of thinking about ethical practice (Purdy 2015). Eduardo Kohn (2014) suggests that new ethics emerge because of a shared planetary risk: "Futures, of human and non-human kinds, are increasingly entangled, and interdependent in their mutual uncertainty" (460). Moreover, the Anthropocene both emerges from and produces scientific knowledge that crosscuts disciplinary boundaries within the sciences: "The power of the Anthropocene epoch and the technosphere, as conceptual tools, lies in their insistence on a large-scale, long-term, systemic grasp of phenomena too often siloed into separate disciplines or analyzed as local or short-term concerns" (Edwards 2017, 40).

The idea of causality shifts in conditions of the Anthropocene: "Relational concepts of agency challenge linear understandings of causation" (Head 2016, 75). The notion of interrelated agency is one that comes from Latour's (2014) notion that the Anthropocene shifts autonomy, that is, it creates conditions in which each subject will "share agency with other subjects that have also lost their autonomy" (5). To Latour, the earth can no longer be "objective," and separated, ontologically and epistemologically, from humans; that is, the actions and choices of human are embedded in the material structure and knowledge of earth. As he puts it, "Human action is visible everywhere—in the construction of knowledge as well as in the production of the phenomena those sciences are called to register" (5). Thus, the simplistic or easy binary divisions between nature and manmade, as well as "utter confusion between objects and subjects," begin to dissolve not only in the ecological reality of the Anthropocene but also in the process of what and how humans know (8). Because of this shared and remade interdependencies, the demand becomes to "distribute agency as far and in as differentiated a way as possible" (15). If we imagine Latour's thinking as a way to reimagine earthquakes, then the earth's seismic infrastructure may be thought of as an agentive force in which human activity is also entangled. If the earth's plates are granted agency in this way, then it would also shift how we understand the causes of earthquakes, as the plates could no longer be seen as somehow absolutely separate from human action. Whereas prior to the Anthropocene an earthquake might be seen as a "force of nature," it might now be a different manmade force that implicates humans politically and ethically (Latour 2014, 16).

Latour (2014) is also thinking about the effect of helplessness in late capitalism: "What is really remarkable is that during the last two centuries the very notions of the two natures have exchanged their properties: first nature has entered the Anthropocene where it is hard to distinguish human action from natural forces and which is now full of tipping points, peaks, storms and catastrophes, while only second nature, it seems, has kept the older features of an indifferent, timeless and fully automatic nature governed by a few fundamental and undisputable laws totally foreign to politics and human action" (6). Stefan Helmreich (2009) helps us out with this distinction between first nature and second nature, which comes from Hegel where the former has a sense of nature as given and law-bound while the latter is the human-shaped aspect of nature which includes social, political, and economic life. In part, what he means is that we are witnessing the complete naturalization of the economy such that it seems to be a self-regulating organism who abides laws that are beyond human control: "The laws of capitalism become . . . more transcendent than those of geophysics and geo-biology" (Latour 2014, 7).

Raffnsøe (2016) likens the Anthropocene to another "Copernican turn" in which humans find themselves as defining the world order: "In an epoch characterized by the over-arching importance of the human species in a number of respects . . . it might be fruitful to picture human beings as situated beings within a topography and examine humans as beings in the middle of the world, on the verge of affecting and redefining themselves" (61). Raffnsøe advocates a decentering of this radical vision in order to appreciate human activity and its consequences (62). Kath Weston (2017) also suggests that part of the radical shift means that how we encounter the world with our bodies can also confuse or distort understandings. When climate change was understood in North America as "global warming," citizens would often doubt cold weather as indicating counterevidence to a scientific consensus. Weston takes seriously how scientific knowledge has often relied on sensory knowledge or, as she calls it, "embodied empiricism" (131).

The notion of the Anthropocene also shifts the idea of responsibility for natural disaster such as hurricanes and earthquakes (Barrios 2017). In Bill McGuire's *Waking the Giant* (2012), he suggests "an increase in earthquake activity is accepted as one possible outcome of the changing climate educing a response from the Earth's interior" (x). McGuire (2012) suggests that "the melting of ice sheets leads to a combination of spectacular uplift and a period of increased earthquake activity as tectonic strain that has accrued on faults beneath and adjacent to the ice sheets is released" (133). One theory suggests that the greater weight of melted water may also contribute to increased seismic activity (138). Now while this is important to recognize the actual causal relationship between climate change and earthquakes, it is also vital to highlight the conceptual shift, the awareness of man-made disaster as an idea, as an available way of making sense of the world.

Though the Anthropocene has not been adopted as an official geological term, it has been used widely in academic research (Chakrabarty 2012; Sayre 2012; Bonneuil 2016; Mathews 2018), as well as in the media (Baraniuk 2017; Roberts 2017; Yang 2017). One can also find media discussion of the *Antropocene*, the Italian term, in Italian newspapers and media (Conversi and Moreno 2017; "I minerali" 2017; Perasso 2015). Why is this significant? The existence of the category becomes a rough way to track when the idea emerged that humans were shaping the earth's climate and ecosystem in dramatic ways. While they may be emic, which is important, there is also a much broader cultural ethos of understanding the environment in new ways, thus an epistemological shift regardless of whether the actual term is adopted or not. One could, alternatively, chart this history by examining environmental movements and a broad recognition that human society—from its agricultural and meat consumption to its emission of greenhouse gases to its waste—was dramatically polluting and poisoning the earth. But the notion of the Anthropocene makes clear that this division tilts from an understanding that humans influence planetary systems to one in which they are a central instigator and cause. Leonard Berberi's (2015) article in the widely circulating newspaper *Corriere della Sera* was titled "Anthropocene, the Year Man Changed the Planet" (*Antropocene, l'anno in cui l'uomo cambiò il pianeta*). Similarly, Elena Dusi's (2016) headline also positions humans in an unequivocal way as controlling the fate of the earth: "Too strong the trace of mankind. Welcome to the Anthropocene" (*Troppo forte la traccia dell'uomo. Benvenuti nell'era dell'antropocene*).

Pope Francis's Encyclical

The powerful influence of Pope Francis in Italy also makes an authoritative plea to Italians to care for the earth, and recognize humanity's role in its destruction. His message further circulates the idea that humans' responsibility in environmental destruction is essential, which also represents a significant agreement between the Vatican and the global scientific consensus. In Pope Francis's (2015) encyclical *Laudato Si' of the Holy Father Francis on Care for Our Common Home*, he laments the disrespect and abuse of the planet: "This sister now cries out to us because of the harm we have inflicted on her by our irresponsible use and abuse of the goods with which God has endowed her" (3). Thus, the text begins with an immediate and clear message of earthly disaster as directly tied to human action. To Pope Francis, the earth is referenced as "our sister, Mother Nature," "our common home," and "sister earth" (3, 4, 39). He attributes the increase in waste and pollution to the "throwaway culture" (Francis 2015, 17). He affirms, "A very solid scientific consensus indicates that we are presently witnessing a disturbing warming of the climatic system. . . . Humanity is called to recognize the need for changes of lifestyle, production and consumption, in order to combat this warming or at least the human causes which produce or aggravate it" (Francis 2015, 18–19). He goes on to detail that of the various factors that have contributed to climate change, including greenhouse gases released "mainly as a result of human activity" (Francis 2015, 19). Pope Francis's message is that human activity is a central culprit, which is actually a quite forceful affirmation not only of climate change but of the Anthropocene.

Pope Francis reiterates this theme of human action: "The modification of nature for useful purposes has distinguished the human family from the beginning" and "it was a matter of receiving what nature itself allowed, as if from its own hand. Now, by contrast, we are the ones to lay our hands on things, attempting to extract everything possible from them while frequently ignoring or forgetting the reality in front of us" (Francis 2015, 76, 79). Here he seems to indirectly reference a key aspect in the Anthropocene debate, which is to say, the notion that humans have always intervened and manipulated nature, and, in turn, the climate, and thus when the Anthropocene began is contested. Yet he also distinguishes the current period as a shift toward increased extraction and use, which aligns with scholars who date the Anthropocene to the Industrial Revolution. He offers a rather nuanced understanding of culture as a way to suggest there are no hard lines between human and nature: "Culture is more than what we have inherited from the past; it is also, and above all, a living, dynamic and participatory present reality, which cannot be excluded as we rethink the relationship between human beings and the environment" (Francis 2015, 108).

Pope Francis reflects on how a biblical understanding of nature has become distorted and self-serving: "We must forcefully reject the notion that our being created in God's image and given dominion over the earth justifies absolute domination over other creatures" (Francis 2015, 49). Moreover, by positioning the environment as a shared resource of humanity, he also endows the correction of ecological wrongs as a moral duty: "The natural environment is a collective good, the patrimony of all humanity and the responsibility of everyone" (Francis 2015, 70).

In the encyclical Pope Francis adopts what Lynda Walsh calls a "prophetic ethos" in which he both reasserts the core values of the church and speaks about scientific knowledge: "Doomsday predications can no longer be met with irony or disdain. We may well be leaving to coming generations debris, desolation and filth" (Francis 2015, 119). For Walsh, such predictions are made only by certain truth-tellers or "scientific advisers" (ix), which may include various kinds of scientific experts, not only research scientists. Pope Francis is known for having a background in chemistry, but his authority as the head of the Catholic Church endows him with a unique kind of authority that, as

Walsh also discusses, blends both religious and scientific knowledge. In other words, Walsh rejects a linear and binary shift between the religious authorities and, later, secular experts on science and instead, as Pope Francis shows us, suggests that such a distinction too easily misses how notions of spirituality represent a part of some scientific advisers, depending on the cultural and historical context.

Referencing his previous work *The Joy of the Gospel*, Pope Francis concludes with a moral imperative: "The gravity of the ecological crisis demands that we all look to the common good, embarking on a path of dialogue which requires patience, self-discipline and generosity, always keeping in mind that 'realities are greater than ideas'" (Francis 2015, 148). Especially striking is his call to embrace reality, as an implicit command to take seriously a single and undeniable ecological materiality. Let us recall that he issued this encyclical at a time when climate change deniers and antiscience rhetoric abound, which is a persistent discourse that refuses, distorts, and denies the existent planetary realities. Pope Francis's words recognize this important truth of planetary risk and disaster.

The Earthquake Hunter and "L'Aquila's Cassandra"

Part of the discourse of the Anthropocene directly shapes how earthquakes are understood, as potentially resulting in an increase in seismic activity (McGuire 2012). But this marked shift in understanding the planet has also produced a greater skepticism regarding scientific research, specifically toward an ability to predict ecological disaster and crisis.[3] Let us now return to the days before the 2009 earthquake in L'Aquila in order to trace how this epistemic uncertainty fueled the credibility of alternative pseudoscientific actors.

A scientific technician at the National Physical Laboratory of Gran Sasso named Giampaolo Giuliani was trying to predict earthquakes using radiometer stations in the area that he had privately funded and assembled. Giuliani has worked as a research technician in physics and astrophysics since 1971, though he does not himself hold a college or graduate degree (Mauri 2012). The field of seismology has been familiar with the correlation between earthquakes and radon since the 1980s, but radon gas emissions have not been seen as a sufficient precursor to reliable forecasting, especially as not every quake co-occurs with radon emissions (Mauri 2012). Moreover, his method of measuring radon gas was not viewed as credible by Italy's National Institute of Geophysics and Volcanology (INGV) and the Civil Protection Department (Dipartimento della Protezione Civile) (Yeo 2014, 407). Giuliani had been denied funding to study radon in 2003 and 2006, which was deemed "at a very low level, from a scientific point of view" (Dollar 2010).

Rising levels of radon gas convinced Giuliani of a coming earthquake, and on March 27, 2009, he predicted a small tremor in Aquila measuring 2.3 in magnitude that occurred on March 28 (Messora 2009). Then, on the 28th, he had predicted an earthquake in the nearby city of Sulmona, and the mayor of the town warned the population by way of loudspeaker and van and, according to Giuliani, exaggerated the threat he had predicted. No major quake came. Three days later Giuliani was put under a gagging injunction and removed the Internet warnings he had issued (Sample 2009; Yeo 2014, 407). Giuliani insists he predicted only a tremor in Sulmona and that reports he predicted a major earthquake were designed to discredit him ("Sentenza L'Aquila" 2009). On the night before the earthquake, April 5, Giuliani reported to have noticed "anomalous activity" on his radon machines and a potential "catastrophic situation" about which he said, "I saw a disastrous event mounting and I didn't know what to do" (Messora 2009). Though the gag order prevented him from public declarations, he did warn friends and family and later wondered "whether they could have saved many more people" (Messora 2009). However, many sources suggest he did not predict L'Aquila, and they counter his narrative at having done so (Mauri 2012). To the scientists and members of Civil Protection who silenced him, he said, "Earthquakes cannot be predicted, they say. Liars" (Marco 2009).

Soon after the actual earthquake struck on April 6, some interpreted his Sulmona warnings as actually predicting Aquila and, by extension, that the gag order had cost lives. Less than a week after the tragedy, Giuliani was receiving international attention and became known as the Earthquake Hunter (*il Cacciatore di Terremoti*) and the "Cassandra of the Abruzzese apocalypse," while others said he merely had a "Cassandra complex" (Messora 2009; Marco 2009; Dalla Casa 2012). The Cassandra reference is particularly of note, as it refers to the Greek legend of Cassandra, who was given the curse of foresight by Apollo yet was considered mad and never believed by the Greeks. She was said to have prophesied the destruction of Troy, the Trojan horse, and Agamemnon's death, despite never convincing the public of her forecasts. Some scientists have said that the trial has made earthquake scientists into Cassandras because they'll be forced to overestimate and forecast danger, knowing the public is not liable to trust them (Kolbert 2015).

Giuliani was also sometimes referred to as a "scientist" yet worked as a lab technician (Sample 2009). Working in the acclaimed Gran Sasso laboratory lent credence to his warnings and allowed for a first media mutation: representing him not as technician but as scientist (Sample 2009). Giuliani's prediction also matters because the Risk Commission's fateful March 31 meeting and press conference were, in fact, to address the rising alarm in the city. Giuliani's predictions in Sulmona had created a climate of panic and anxiety in the region to which the Major Risk Commission responded.[4] In particular, their testimony clarified that neither swarms nor radon spikes were scientifically verified as forecasting earthquakes (Kolbert 2015). Marouf Hasian, Jr., Nicholas Paliewicz, and Robert Gehl (2014) have argued that "manufacturing of his ethos involved a great deal of boundary work aimed at placing him outside the parameters of accepted science" (566). After the earthquake, the scientific community held that anything he did predict was a "lucky guess," and he earned a reputation among local citizens as a caring if misguided independent researcher (Hasian et al. 2014, 566). They also suggested that the Major Risk Commission enhanced its own authority, "magnifying" its "own social agency," by talking about the failures of predicting earthquakes by radon and describing Giuliani's work as "at a very low level, from a scientific point of view" (Hasian et al. 2014, 566). Further, public reaction to Giuliani was manipulated to confirm claims that Italian publics were illiterate and naïve about scientific prediction.

Nico, who also worked at Gran Sasso, had this to say of Giuliani:

> He was a technician; he calls himself a scientist because he does science, whatever, but it was important that he worked for Gran Sasso. He was a researcher, went to university, did his doctorate, and then started doing research on radon. I don't think he came out with scientific results; he used a method that other scientists have discredited, I mean, official science. He used that method. I mean, it's not the fact that there are correlations between earthquakes and radon emissions, that's been known since the seventies, this is known. The problem is defining correlation. Radon is an element, a noble radioactive gas and so it comes from a chain reaction of Uranium 238, which are heavy elements. Radon is an element distributed in the ground in an almost uniform way, right? And on earth there's uranium, there's radon, and

radon respects other elements on the chain: it evaporates and comes out on the surface. . . .

I'm reminded of the film *Dances with Wolves* when, at a certain point, there was an abandoned field, and he moved this whole wall from where it was because there were buffalo. The terrestrial crust is a kind of sponge. There are many reasons to have radon emissions, after an earthquake as well as before an earthquake. But you can't say that every time there's a radon emission, there's an earthquake. Giuliani doesn't predict it every time, but many times when he gave the alert, it never happened. It should be studied in the right environment. But you need to use the scientific method, not a semi-scientific one. . . . He had alerted people that his instruments had detected a radius of 50 kilometers, which included the town of Sulmona. And he said there would be an earthquake. He made this news, exploiting the occasion to break out from the game of not being listened to; he'd been considered a charlatan.

Noelle. And people listened to him?

Nico. Yes, a lot, so, very much! He was more famous than the guys in L'Aquila. . . . Oftentimes the problem is not letting one's ego go and the mania of being a hero [*protagonismo*]. In science, there are numerous heroes, but science isn't made of heroes, it's made of facts. . . . The world is full of behind-ism [*dietrologia*], those that want to call out disinformation, they are like occult theorists with ideas about what is hidden.

Nico's insights here on "behind-ism" help us understand why a broader epistemological shift in natural or manmade disaster, together with a rising scientific skepticism, weakens public ability to trust information and experts, especially on imminent ecological crisis.[5] While Nico still expresses a kind of distinction between science as factual and behind-ism as fictional, the latter seems increasingly contingent on and necessary to outline and differentiate scientific knowledge.

Conspiracies and the Man-made Natural Disaster

Let us reflect on co-occurrence of epistemological shifts which shape one another and how people understand disaster: the scholarly naming of the

Anthropocene, the cultural awareness of human-driven climate change (not necessarily called anthropocenic), and conspiracy theories regarding man-made terrestrial disaster. In particular, and as a way to more fully understand the rise of Giuliani, there has been the rise of the concept of man-made artificial earthquakes, a growing rumor in Italy about deliberate and intentional environmental disaster. These conspiracy theories indicate a weakening in public trust in scientific expertise and knowledge and an uptake, if indirect, of anthropocenic causality.

I first heard about man-made earthquakes from an old friend of mine from Padua in 2011, Roberto, a geometer who worked in telecommunications infrastructures. I was in the middle of explaining a new research project on the 2009 L'Aquila earthquake and trial when he said something that stunned me: the 2011 Japanese earthquake and perhaps many others, he said, were actually caused by secret warming devices that heated the earth's core and caused plates to shift. He was not talking about fracking, but rather massive aerial machines. He believed the earthquake was created by the United States. Since then, I've tracked this theory in conversation and online in Facebook stories titled "How to Create Artificial Earthquakes" detailing the microwave-like underground waves that gave rise to dangerous earthly tremors.[6] In 2016, after another round of small quakes between the regions of Umbria and Marche, the Italian newspaper *L'Espresso* reported that "disinformation" and behind-ism were behind new rumors about the underlying cause of the quakes (Grandinetti 2016). The first theory was that the government offers public funding only when earthquakes occur with a magnitude above 6.0, such that government's negligence became the theory when an initial magnitude of 6.2 became 5.9. Meanwhile, the following alert was circulating on the text messaging program Whatsapp, erroneously credited to Civil Protection agencies: "Pack your bags and go because strong quakes are predicted over the next few hours" (Grandinetti 2016). The paper cited a popular tweet: "To end with the popular beliefs and 'an earthquake mood [*l'aria di terremoto*]': the muggy heat lately was really strange, the gray sky and hot wind. . . . I don't know if it's connected but it makes me think. #Terremoto."

Part of what we see here are clashes of information, shifting messages that circulate from different kinds of authorities, knowledge, and senses: government agencies, sensory responses to the atmosphere, with a keen sense of how interests might sway truth, particularly with the theory that the government might underreport magnitude. Several popular books are dedicated to unveiling or discrediting this kind of subterranean theory of the world, including *Misinformation: A Guide to Information Society and Credulousness* (Quattrociocchi and Vicini 2016), *Stories of Ordinary Falsehood: Urban Legends, Fake News, Lies; The Macroscopic Falsehoods Told by News, TV and Internet* (Toselli 2004), *The Era of Post-truth: Media and Populism from Brexit to Trump* (Cosentino 2017), and *Disinformazia: Communication in a Time of Social Media* (Nicodemo 2018).

Another set of theories are known as "chemtrails" (*scie chimiche*), which posit that commercial airplanes deposit chemical substances, such as barium, aluminum oxide, thorium, and other heavy metals, in the air in order to malnourish and genetically alter populations. The idea of chemtrails actually began in the United States and because the United States Air Force published a report on weather modification. An Italian website dedicated to chemtrails proposes that "those long, wide, and persistent trails cannot be condensation" (Scie Chimiche 2007). Moreover, the site suggests a relationship with climate change: "We arrive at the notion of the strange correlation between airplane trails and the climate change in course. People began saying that the trails influence Earth's climate, then passed to affirming that they are the cause of global warming, to then conclude that they have an effect on the imminent cataclysm for which humankind is 90% responsible" (Scie Chimiche 2007).

Many who believe in chemtrails refer to the American HAARP [High Frequency Active Auroral Research Program], an ionospheric research project, as proof of chemtrails, as the project develops electromagnetic communication. The HAARP project has a central Alaska–based research site which includes an Ionospheric Research Instrument (IRI) capable of transmitting high-powered radio frequencies. According to the chemtrails Italian site,

> A similar technology would be capable of raising and warming vast areas of the ionosphere and emitting elevated quantities of energy; bouncing electromagnetic waves capable of penetrating anything alive or dead, in predetermined sites on Earth. This type of stimulation would cause molecular modifications in the ionosphere, which would bring devastating consequences to the climate in the hit regions. In short, HAARP, being capable of provoking intentional climate change, could be used as a military weapon. Moreover, the waves reflect on the Earth's surface, [and] because of their elevated

> intensity and capacity for penetration, would also be capable of manipulating and disaggregating human mental processes. (Scie Chimiche 2007)

In this horrifying scenario, man-made machines and technology can "penetrate" bodies, even dead bodies, and enact mind control, in invisible and unexpected ways.

The website has a special section, "Disinformation," which the authors define as "the conscious diffusion of information purposely erroneous and distorted in order to influence public opinion on a particular subject" (Scie Chimiche 2007). The site owners consider the following under the rubric of disinformation: reassuring references to airplane trails as normal and safe, calls of chemtrails as "plots" or "conspiracy theories," websites seeking to decry chemtrails as paranoid and nonexistent, and television programs and journalism as ignoring the topic. Moreover, they detail how the mind might be induced to misinterpret an everyday news transmission:

> The disinformation most subtle is hidden behind subliminal messages. . . . For example, if there were a scene from a film that showed long white trails in the sky, probably in that moment, we would recognize that detail, as long as our attention was not diverted. Our brains, however, are capable of assimilating that image. If then, say, we see a similar image in another film, then a third film, then a commercial, then in a second commercial, then in a cartoon, then in another animated film, then on the news, then on a second news report, then in a public pamphlet in our city, then in a second pamphlet, then in a magazine article on gardening, then in a second article, then in a third, then between the pages of a newspaper as the backdrop of an appliance ad, then in a second paper, and so on. At that point, most likely, our brain would accept as real that the sky is full of long white trails and when we saw one like it live then we wouldn't be minimally shocked (Scie Chimiche 2007).

The "robust" disinformation works by populating and disseminating the image across the informational sphere. It normalizes the image of the white emissions in the sky by purposeful repetition across a variety of media. Here, the networked media are imagined as a tightly orchestrated and normalizing apparatus, manipulating the slow and continual exposure to harmful agents and reducing the fear and curiosity in civilians. This idea of agents working purposively to undermine humans' responses of fear and not rely on their own sense of danger is echoed in Aquilan citizens, in accounts of the time leading up to the earthquake.

Fredric Jameson (1995) argues that conspiracy and paranoia in late capitalism are no longer about privacy but about the heightened amount of corporate power: "[T]he 'conspiratorial text,' which, whatever other message it emits or implies, may also be taken to constitute an unconscious, collective effort at trying to figure out where we are and what landscapes and forces confront us in a late twentieth century whose abominations are heightened by their concealment and their bureaucratic impersonality" (Jameson 1995, 3). The "concealment" is about a shift in the locus of knowledge to submerged government centers and loci which have become increasingly harder to discern and place.

Chemtrails are not the only conspiracy circulating that positions scientists as controlling the climate and creating disasters. Scientists are referred to as "scientist-puppetmasters" in a *Time* story about the creation of rain storms in Abu Dhabi (Sanburn 2011). The $11 million project was able to bring ionizers to the desert that "produce charged particles," which attract dust and form a cloud after rising in the desert temperatures. In 2009, China reportedly used this same technique, called cloud seeding, in order to take measures against a drought and ended up creating a snowstorm (Moseman 2009). Without delving into the history of cloud seeding, it seems significant that it has been around since 1946 and was apparently used by the United States to modify hurricanes in the 1960s in a well-titled "Project Stormfury." In Italy, reports circulate that cloud seeding [*inseminazione delle nubi*] is designed to increase rainfall and has been used most often by Russia and China ("Cloud-seeding e chemtrails" 2008).

The point is this: what sounds supernatural or conspiratorial may in fact be a savvy and perhaps even hyperrational analysis of human complicity in disaster and environmental change. In this emerging worldview, climate and so, too, natural disasters become premeditated homicidal agentive acts, politically motivated. The "artificial earthquake" reimagines earthquakes as the pinnacle of scientific knowledge, deliberate deployments of exquisite scientific knowledge that masquerade themselves as "natural disaster." It does not fall prey to seeing human action or science as disinterested or objective. It thus unlocks the secret as to what was so threatening about the trial and why the decision was overturned. The "artificial earthquake" reimagines earthquakes as politically coerced and planted.

But legal systems and cultural beliefs across the globe are just adapting to this new paradigm, as are studies on the matter. We must reckon with what the Anthropocene means for how we understand human responsibility, legally, politically, and culturally. Integrating this responsibility gets especially tricky given that our ecological environment is shared and is a collective good, while there exist profound global inequalities between the contributors, knowledge producers, and policy designers who shape ecological destruction and those who become its victims. As a concept, the Anthropocene points to collective human action on a massive scale, yet in Aquila, for instance, the responsibility is atomized within a small set of scientists giving misinformation. Aquila's trial was an attempt to take scientific misinformation to task and may have happened, in part, because we have a greater cultural awareness that humans, not just nature or God, are responsible for human death that results from natural disasters. There are significant implications here for a way to think about the ecological conditions, scientific knowledge, and law: could we begin to hold climate-change-denying scientists responsible for environmental damage or loss? We live in an age where we will see increasing natural disasters and new techniques for legal and political management: Aquila's earthquake trial and surrounding conspiracy theories from Giuliani the Cassandra to chemtrails are evidence that notions of individual accountability with regard to natural disasters are deeply unsettled.

Expert Reassurance as Behind-ism

Within this culture of behind-ism pieces of information are suspicious because every piece of information presumably has secondary or tertiary meanings. It is a world in which the grand interlocking networks of knowledge and information become fragile and fraught, in which the government disseminates information according to its interests, and in which the state and media together erode citizens' capacity to distinguish truth from falsehood. It is here, too, that we see how the faith and knowledge in scientific discourse also erode.

On a sunny day in June, I met with Clara, a local resident and architect, and as we sat in her studio, she spoke with great wisdom about how reassurances overtake and confuse basic human instincts of fear and common sense, which have been culturally habituated especially in seismic areas like L'Aquila:

Clara. I work in L'Aquila in restructuring, since the earthquake; I've always been in restoration but afterward I dedicated my work to the reconstruction. In the whole period before the shocks, for about four or five months, I always had this awareness that the city was not secure. I found proof of this after a few shocks. I did some building checks and saw the situation in certain buildings. If anything, I always tried to sensitize people to pay attention, to leave their homes. I had one unusual moment when, the week before the earthquake, there was a significant shock. I was at a meeting for the local order of architects, and there was a shock. We called the fire department and the firemen came. I told them that I saw some fracturing in the stairwell and in some areas of the apartments, in this beautiful, historic building where we had the architect meeting. They seemed dangerous to me, I remember. So then they assured me, and then told me I shouldn't be alarming the building residents in this way, which was true in some sense because I don't have a structural background, but it's also true that I've been an architect who has worked in the field of restoration for thirty-five years and so I have a certain knowledge of old buildings. I can see certain arrangements, things I understand because my experience lets me see them. I can't technically explain the exact moment of rupture or when an event might happen but I can understand vulnerability. I have to say that in this whole thing in L'Aquila this aspect of safety has been strongly undervalued because if only we had been allowed to do what all of us have always done. Ever since we were little, and I remember people would say that when there are shocks you had to leave the building or put yourself under some secure structure, like arches . . . or the door frame, so you had to have a secure place to put yourself. Yet here both the technicians and the specialists fundamentally assured us—they assured everyone—and this was emblematic.

Clara is witness to this strong tension between embodied and experiential knowledge which was clashing with repeated expert advice. She reflects on this in terms of her professional capacity to identify weakness and vulnerable points in a building but also draws from her childhood. It is if she is asking: What happened to that embodied knowledge of the danger and risk of shocks and earthquakes? Why did our own knowing not override the so-called expert knowledge? First, the fire department did not take her evaluation of this at-risk building seriously; then, the long-held practice of exiting buildings seemed to lose force because of the more generalized public assurances. Moreover, while she identifies the fire department as one site of assurances, she also generalizes to "specialists," which refers also to Civil Protection

Services and a broader network of scientists and seismologists who shifted the discourse toward one of safety and assurance. Finally, it is especially important to note that she suggests she was told not to "alarm" the residents. Here, Giuliani's predictions of the shocks and public panic produced a reactionary climate, in which claims of danger and risk were silenced, with the protection of public order remaining, an underlying motivation to hush anyone, even experienced architects, from inciting or agitating residents. The cost, as Clara tells us, was great because this push made for overriding the practices that most residents had been habituated to since childhood.

Clara also described another event that illustrated this same push toward safety over risk, at all costs, in ways that ironically become nonscientific, a wholesale and irrational denial of empirical risk:

> This happened to me, me and my sister, who is also an architect but ten years younger than me. . . . She had a beautiful home on Via Marciano and two days before the earthquake there was an important shock during the day. We were both in the studio on Via Cascina and we were called by Angela's children because a big hole opened up in her living room in this beautiful old building. So we went to see it, and Angela said, "Let's call the fire department." I said you have to leave the house; go to where we have the studio; we have our studio in an antique building but we restructured it. So I didn't want to put my hand to the fire but you see nothing happened at our studio. She said ok, but let's hear from the fire department. They looked around, took a turn around the place, and said "There's no problem here. Don't worry." I told them, "But Engineer, are you sure?" And they told me, "Yes, don't worry, Architect, there would have to really be a terrible shock to make this building collapse." I saw this very gentleman the day after the earthquake, and he looked at me and said, "The terrible shock came." I told him, "Engineer, my sister and I work on the other side of town; she should have stayed there." "Certainly," he said, "I would have stayed there too."
>
> So the real atrocity in L'Aquila was—how can I put it?—reassuring a population that had been used to being afraid of earthquakes. All of us were used to being afraid because they had taught us since we were little that we had to fear earthquakes; an earthquake is not a joke, it's not something in which there is no fear because you can be buried under an earthquake. And my sister didn't die only because there the rooms that fell were in another part of the apartment than where they were sleeping. And a lot of deaths were only because people trusted in these reassurances. Those that were saved from col-

> lapse were those who left; they did what their grandparents told them to do. There were a few reasons to assure people, but certainly not worth the deaths. It was a shocking way to think that these things don't happen. Things happen. Where there is no safety, you have to do it in a way that isn't urgent. That wasn't done here, I mean, I think that not all three hundred and something deaths could have saved themselves but a good part could have been saved. Because there were people that didn't leave their homes because they were reassured. For example, I have friends who lost their children because they didn't leave that night because they were reassured. A dear friend of mine lost his wife and daughter; his son didn't die because he was on a school trip. So what happened here was a maneuver of insanity [*operazione demenziale*], plus, panic would not have broken out for the simple fact that people were already used to doing this.

Here, Clara tells us of yet another moment of tension when her professional experience and sense of danger were undermined by the fire department. In between her curt but respectful exchange—note her use of proper titles of "Engineer" and "Architect"—were once again disciplining and dismissing her real concern. We might liken what Clara describes as a "maneuver of insanity" to a kind of mass public gaslighting, the idea of inducing a doubt of one's sanity in someone else. Here, even a rupture in the building at her sister's home could not elicit agreement that her sister should not reside there further. Until, that is, after the earthquake. While her own expertise was dismissed and her doubts about safety were dismissed, so too was the power of these "reassurances" to override instinct, habit, embodied knowledge, and fear.

We have to appreciate the danger of these quasi-scientific reassurances—in the sense that they were usually expert based, coming from the fire department or local experts—in rewiring a public to respond differently to emergency and danger. This echoes the trial witness who likened the role of expert knowledge as a kind of "anesthetization" to danger. Now, one may argue that this notion of expert anesthetization was a local, available discourse, one tied to and further disseminated by the trial coverage, for other educated citizens like Clara to adopt and reproduce. However, I see anesthetization as a structure of feeling, that is, if we consider it a kind of epistemological anesthetization in which this mass-mediated attenuation of risk and bolstering of public safety eroded these deeper instincts and familiar practices. The stakes of this fear-blunted and erroneously safeguarded way of seeing were deadly: Clara is persuaded that this shift in apprehending danger was the lynchpin in the high Aquilan death toll.

I asked her, "Where did this culture of reassurance begin? From whom?" She said,

> It was really Civil Protection. It was at a central level, if we consider that the same agency, Civil Protection, had asked to work on all of the buildings at risk in the city. Therefore, the problem came in that this publication was done by an organization that had analyzed all the buildings and they were evaluated on every criterion so they *knew*. Our luck was that it happened at night because if it had happened during the day there would have been even more deaths because all the public buildings fell; the schools fell. We would have erased a generation because all the schools came down.

She clarified that this study had come out after the earthquake through the trial, but it was shocking that the risk of full and widespread building destruction was known. She then reflected, "I don't know whether it's from craziness [*demenziale*] or from trickery [*truffaldino*]." The anticipatory Civil Protection study of risk heightened this sense of a "centralized" governmental effort to bolster public reassurances at all costs; moreover, it too is a kind of affirmed "behind-ism" insofar as it laid bare the underlying administrative motive, which was to convey the impossibility of full evacuation of a city in which the majority of buildings were at risk of collapse. Here, in these logics of the conspiratorial thinking, it becomes especially hard to decipher intent. Were it merely madness, Clara implies that the governing institutions had no concerted and coordinated effort to harm or undermine citizens. On the other hand, were it "trickery," this implies an underlying intent at public manipulation. What is important is that Clara is wavering between these possibilities, effectively embracing neither a fully conspiratorial view nor one of full trust.

Darkness on the Surface

Meanwhile, it was time to return from the dark belly of the earth as Giacomo and I set about back to his car, to funnel, once again, toward the light and the surface of the earth. I was eager to return to sunlight, yet he seemed somewhat indifferent to the growing light. It reminded me of what he had told me earlier that morning, of a memory of a darkness in postearthquake L'Aquila that truly impacted him:

> I remember, I was alone and went to take a walk around the city. The historic center was still closed and I went through the barriers and took a walk around the city. Knowing it was the day of the Feast of Celestinian Forgiveness [*Festa della Perdonanza Celestiniana*], hearing the sounds of the festival in the distance, and feeling a freezing coldness in the city that had always lived that festival. To me it seemed like something that . . . [his voice faded out]. I get goosebumps thinking about it.
>
> There was this coldness, coldness, a strange chill, almost death-like, in these abandoned streets. And it was real, not some kind of emotional response, but really true. Even now, when I pass there's this humid coldness, where there were people living and there aren't anymore. It's all uninhabited and deserted, and there's this coldness, a mortal chill. The city just gives me these incredible impressions. And then the darkness! Darkness, pitch black. Not even a streetlight. Being in the red zone there weren't any lights so there was this darkness. To walk, I had to go by memory.

In this moment of darkness and loss, a city abandoned and destroyed, Giacomo seemed to return to a place of knowing that has been assaulted within this particular regime of epistemological anesthetization, and find, once again, that his own memory was stronger than darkness. And we have analyzed Aquila's trial in terms of Italy's former prime minister Berlusconi's culture of disinformation. But, as I learned from Giacomo, sometimes you have to change scales: if the burrowing underneath a mountain lets us sense the black hole thousands of light years away, so too that mountain's quaking might make more sense in light of planetary paradigms.

Conclusion

Mirrored Window World

The truth is that a human is just a brief algorithm.

—*Westworld*

Like the witch in *Snow White*, people in our technological society look into the mirror of their clean little society, dominated by capital, and yet they see beauty, extreme beauty.

—Beppe Grillo, *Il Blog di Beppe Grillo*

Google knows what you search for and can infer not only what kind of news you read on a Sunday morning and what kind of movies you prefer on a Friday night but also what kind of porn you would probably like to gawk at on Saturday night after all the bars have closed.

—Colin Koopman

Before a crowd of parishioners, the minister unveils a mirror, hailing it as a symbol for the new religion he's created, "selfism" (*lo ionismo*), with the slogan, "You are your own god" (*Tu sei il tuo dio*) (*Io C'é* 2018). In the 2018 film *I Exist* (*Io C'é*), Massimo Alberti owns a bed and breakfast in Rome and seeks a way to boost profit. After admiring the bustling tax-free business at the neighboring Catholic hostel, he decides to form a religion of his own to avoid paying taxes. He gets together with friends who pick and choose among practices of world religion in order to invent his own religion: selfism (*lo ionismo*). Selfism, they declare, will have few rules: no special clothing or hats and no dietary restrictions. But he appears to wonder, what could take the place of the crucifix? If the religion's god is oneself, then it follows that the mirror is the perfect symbol. Besides, Massimo jokes, mirrors are plentiful and cheap. The new religion quickly gains several followers whose services include sitting in front of the mirror and talking to oneself in order to navigate life as their own god. Yet things quickly go awry. One new devotee is a woman facing cancer who tells him she may forgo treatment and dedicate herself solely to "selfism" prayer. His plan backfires more spectacularly when his followers become such fervent believers that even Massimo's confession that he made up the religion is reframed as the prophet's "test." In his frustration, Massimo smashes and destroys the interior of their chapel. What started as Massimo's cunning vision of tax savings ends up with his imprisonment for the destruction of sacred property. The film leaves viewers with a final image of his parishioners sitting faithfully at a service, with the mirror at the altar, and a new painting of their now persecuted prophet, Massimo.

On one level, *Io C'é* is a narrative about how a belief system can surpass its own creator. It works as a satire of tax evasion for everyday Italians and humorously plays on the financial benefits of the Catholic Church. But it can also be read as a commentary on personalization trend and disinformation in the twenty-first century. The mirror becomes the double symbol of disinformation and falsity yet also parishioner belief and devotion. The icon transports individual attention from the three-dimensional to the two-dimensional world, from material people and bodies to immaterial and twisted reflections. The mirror holds seemingly magical power over people: the more they gaze into their own eyes, the more empowered they feel; the more they see their own likeness, the more they feel as if they understand the material world. As gods of themselves, the devotees are literally self-sacralized. It is their own disorientation, deepened by indulging their narcissism and obfuscating their gaze, that blinds them from seeing the actual material reality.

Of course, the faith in the new religion relies not on magic but rather on mystification: film viewers see the careful and greedy deliberations to create selfism, but parishioners do not. Instead, they enter the new sacred space without knowledge of how or why the mirrored altar was created, believing it to be in their interest and for their benefit. Thus, the mirror serves as a kind of double illusion: it is a fake religion icon, created for someone else's financial gain, not individual enlightenment. Plus, it reroutes the devotee endlessly back to oneself, even though these worshiping lookers experience the process as open-ended, transporting them beyond the room and gaining new knowledge about the world.

Knowledge and Power

The Truth Society has examined how what we know about the world shapes who we think should rule it. Its central preoccupation has been examining particular knowledge practices and their political outcomes, and new political practices and their effects on knowledge. We began with mediatization—shaping political facts and knowledge into glossy and sound-bitable nuggets—that gave rise to Berlusconi's right-wing business-outsider political supremacy as well as pro-science skeptics who uphold objective truth. Mediatization emerged from and sustained political cynicism regarding the truth, wrought through over twenty years of former prime minister Silvio Berlusconi's fabrication savvy. Yet because politics became so overpopulated with fictions and spectacle, the fantasy of absolute scientific truth became seen as a corrective to the onslaught of fabrications. At the same time, scientific experts were in unstable and polarizing positions, shifting between overly trusted yet also overly dismissed. Truth, particularly mediatized truth, became so populated by nontruths and spectacle that the fantasy of scientific truth became almost mystified. One vector through these events is how scientific truth and rationality become seemingly absolute yet also radically discounted, both emerging from a historical moment in which the very notion of reliable truth appears scarce.

This terrain of mediatized truth was also a precursor to algorithm society, where most knowledge is filtered through online algorithms, generating user data and feeding it back through customized searches and products. Here, the Five Star Movement rose on a wave of algorithmic society but also the same cynicism and paranoia about mostly televised information that Berlusconian Italy had wrought. Indeed, the innovations of digital democracy of the Five Star Movement may have been fueled because citizens were escaping from the censorship and suppressive rule of the television and radio.[1] That science was both undermined and valued during this period made for mixed outcomes, including strong reinvestment in science that verges on enchantment. For the dice-wielding scientific crusaders of the Committee for the Investigation of Pseudoscientific Claims (Comitato Italiano per il Controllo delle Affermazioni sulle Pseudoscienze, CICAP), the logics of probability have a kind of conversion power to immediately eradicate and excise superstitious belief. Citizens in L'Aquila imagined themselves as anesthetized by science such that their own phenomenological knowledge of place was overwhelmed by "scientific" reassurances. The trial in L'Aquila, in this light, shows us the dangers of adhering to scientific judgment and the juridical attempt to hold scientists legally accountable for their espoused disinformation. The politics of Aquilan pre-earthquake anesthetizing and postcrisis management was also grounded in a different set of epistemological shifts: principally, the widely observed trend of rising "occult" thinking and enchantment. Here we traced how precarious economic and political life gives rise to magical belief and supernatural cultural imaginaries, which, in turn, lead to a kind of enchantment of scientific knowledge production. Finally, the Italian disinformation regime is also set against a new epistemological paradigm-breaker, the Anthropocene, in which human agency becomes the culprit for environmental disaster. Not only are patterns of ecological causality shifted and destabilized, but scientific expertise is decentered, which allows eco-conspiracy theories—specifically, "beneathology" theories of chemtrails and earthquake warming machines—to thrive.

To conclude, then, let us continue to dwell on our methods for information gathering to shape new political forms, specifically examining the political implications of algorithmic processes that become, much like Massimo's church of selfism, a kind of mirror. How does a world of mirrored knowledge shape politics and bolster new political forms? Algorithms, then, represent a key nonhuman agent present in the making of post-truth society and governance. What began as a mass commodification of data has implications for further obscuring knowledge production and predictive logics which, in turn, allow consumer practices to mystify self-knowledge. Yet while the citizenry depends on a predictive and personalized regime, so too does such customization assist in distancing the citizen from understanding these mechanizations and from scrutinizing how forms of governance might scaffold control and discipline on these data. The past decade or so represents a particular nexus of information society in which the customization of data engenders the silo-fication of the individual in what has been called "filter bubbles" (Pariser 2011). Political subjects are rendered especially vulnerable for two central reasons: first, information appears neutral and its self-customization is obscured. Thus, users do not know they are receiving ever-personalized news and media. Second, as filter bubbles grow around individuals, the algorithmic production of data is increasingly mystified and yet, above all, invisible. Perhaps most urgent of all is that this often-ignored structure of Big Data, which has fundamentally changed how we access digital information, has already become the prevailing infrastructure. Moreover, what Nick Couldry (2014) dubs the "myth of Big Data" is "changing the terrain on which all large institutions (including governments) can *claim* to tell us the way things are" (887). Self-replicating information practice gives rise to a kind of absurdist populism, characterized by authoritarian yet ineffective leaders and increasingly polarized political subjects deprived of information outside of what was already precalculated to their tastes and interests of "informational persons" (Koopman 2019).

The Mirror Stage: The Personalization of Digital Life

Time magazine's person of the year 2006 was "You," Internet media thrive on user content, Netflix series marketed to Millennials and Generation Z are titled *You*, and Fitbit wearable devices promise "a healthier you" (Schüll 2016, 321). Various products market the idea of personalization from fragrances to home computer assistants, curated clothing, and food and wine delivery. Algorithms personalize our data, whether known to us or not, as we navigate the Internet: everywhere we look are marketing, knowledge production, and consumerism tailored to and already predesigned to individual patterns of behaviors and preferences: customization capitalism. The first stage of the digital shift was lifestyle marketing in the 1990s which involved "demassification," or marketing toward a unique individual consumer (Leiss et al. 2018). Emerging in the late 1990s to mid-2000s, the movement was also largely driven by enhancing cookies or digital tracking technologies within Internet browsers which could personalize ads to reflect users' unique online browsing (Beck 2015, 129; van Dijck 2014). Digital trackers combine information in web browsers that link different Internet platforms and, via algorithms, deliver information to third parties to then feed personalized ads and links to the user browsing history, with most of this digital calibration unbeknownst to the user (Beck 2015; Koopman 2015). Thus "data capitalism" or "datafication" was underway from even the 1990s when online tracking technologies were commodified as a central part of the information economy (West 2017; van Dijck 2014, 198; Koopman 2019; Navarria 2019).

Yet later iterations of data capitalism went from individualized marketing to customizing entire product designs and services with companies like Birch Box, Stitch Fix, and Winc with home delivery services tailored to individual preferences and desires. Such services principally offer consumers surveys but also use digital tracking via social media and other websites. Another intensification of the demassification trend or personalization trend involves digital technologies that users attach to their bodies. Natasha Schüll (2016) examines the wearables market, the design of wearable tracking technology and the continuous feedback loop that produces immense amounts of new data and also caters to individuals' health: diet and nutrition, movement, and sleep. Schüll suggests there is a new kind of digital self emerging, one encouraged to constantly produce and measure self-referential data: "We are invited to view ourselves as longitudinal databases constantly accruing new content" (325). Notably at stake in this shift in this digital self are the presumed inadequacy and inefficiency of "naked" self-care without digital assistance and augmentation. With all of the personalized consumer schemes, personal data are framed as optimizing the good or service. Customization capitalism depends on user data for marketing and branding but also relies on algorithms to fuel the good or service, and produce a society with user surveillance as a primary economic and social mode, thus fitting alternative names for this assemblage such as "dataveillance," "surveillance capitalism" and "data colonialism" (van Dijck 2014; Zuboff 2015; Couldry and Mejias 2019).

The stakes of algorithm-driven systems are high. Technology theorist John Cheney-Lippold (2011) suggests that as algorithms become increasingly more efficient and more data about a user become available, we find a more intense and highly modulated organization, efficiency, and precision of one's algorithmic identity. He argues: "The algorithm ultimately exercises control over us by harnessing these forces through the creation of relationships between real-world surveillance data and machines capable of making statistically relevant inferences about what that data can mean" (178). Algorithms are indeed getting to know us, perhaps better than we know ourselves. Moreover, as the world becomes more interconnected with more data to calibrate to the individual, the algorithms will actually become more precise and smarter: better able to predict consumer patterns and movements, and thus, better able to reshape production.[2] Hugh Gusterson, in his conclusion to *Life by Algorithms*, "RoboHumans," reflects on the implications of algorithmic systems on our minds: "Computers offer a model of cognition that increasingly shapes our approach to the world, even when computers are not directly involved in information-processing. . . . Human beings are learning to think more like computers" (Gusterson 2019, 10). Gusterson is concerned about how people learn to adapt to algorithmic life but in ways that are fundamentally dysfunctional, such as the American Alt-right gaming Google's auto-complete function such that queries like "Are Muslims . . . ?" would auto-fill with "bad" or "evil" and lure users to their hate sites (43). Another example is modifying résumés to fit keyword searches on LinkedIn or other hiring software.

But in each of these examples there is actually a problem even before citizens begin altering their lives and ways of thinking toward the algorithm, which is the isolating effects of the user's own algorithmic imprint. Yes, Google's auto-complete algorithm comes from millions of users around the globe. But Google is also calibrated to the individual user's past searches, zip code, website, Amazon purchases, and even Facebook click history. Some of these calculations depend on how often a user erases "cookies" or embedded programs of user history and clicks (West 2017); but by and large, the individual's Google experience is a refraction of the individual's own information-gathering. In fact, this is a mirror algorithm.[3] While the user-citizen imagines access to Google or the World Wide Web that looks outward, as a window to new information, it is actually a mirror, always reflecting the user's own atomized history with traces of the user's unique and individualized worldview in ways the user cannot recognize (West 2017). Not only are the machinations of data surveillance invisible to users, but they appear rather "neutral" or "a 'normal' form of social monitoring" (van Dijck 2014, 206).

We are facing a world where even our self-portraits are algorithmically enhanced. Our own faces are reflected back to us in ways that are distorted and based on algorithmic technology: we literally cannot see ourselves. Digital cameras already use algorithms to best render images from different wavelengths of light (Madrigal 2018). However, newer phones are further tweaking these basic algorithms in order to enhance the appearance of the user; what's more, while some apps like FaceApp or "beauty mode" allow for users to select when to use these options, the iPhone XS makes this function the camera default. Thus, the user unknowingly appears enhanced without recognition of the algorithm; that is, all of this occurs "under the hood"—hidden and obscured from user awareness (Madrigal 2018). The potential for commodification is high vis-à-vis facial recognition or even sales of the iPhoneXS, not to mention future scenarios where the phone becomes a surveillance mechanism. The seemingly innocent selfie represents both the surveillance and the new age of mirrored algorithms, with a clear warning: "The global economy is wired up to your face. And it is willing to move heaven and Earth to let you see what you want to see" (Madrigal 2018).

Logically, one can infer that these machine-learning algorithms are based on known facial aesthetics and enhancements: larger eyes, smoother skin, contoured face, highlighted cheeks. Yet they may also—today or in the near future—be calculated based on facial detection software, posted and "liked" selfies on social media, favorited phone photos, even the user's own past selfie uploads. Therefore, the selfie is not only an enhanced self-image but also a reproduction and feedback loop of the individual based on past personal preferences. Mirrored algorithms produce an invisible narcissism, making readers more beautiful to themselves in ways that are produced through and by way of their own unique aesthetic behaviors and preferences that they cannot trace or even recognize. These kinds of prisms engender sensory confusion in that a regular camera, which might produce a more faithful self-portrait, would slowly become seen as fake or wrong, as the person becomes habituated to the person's enhanced image.

Enhanced selfie algorithms are a prime example of mirrored knowledge, algorithms and automated processes that both reflect and reproduce the individual user. A more precise term might be hidden mirrors or even narcissistic algorithms because the user is almost always unaware that such use and information gathering are already prestamped with the user's preferences, history, and data amalgamation.

The Future You

It is not just that mirror algorithms are capitalist systems catered to the user but also that the future is predicated on these individual tastes and preferences. The next step of customization is about predicting what users will do and want in the future. *Dataclysm*, Christian Rudder's best seller, argues that algorithms forecast intimate knowledge from the most seemingly insignificant data. Rudder shares, for example, that our pattern of "likes" on Facebook can predict one's sexual orientation "with eighty-eight per cent accuracy," and with somewhat less accuracy (66 percent) predict whether the Facebook user's parents were divorced. He says, "You know the science is headed to undiscovered country when someone can hear your parents fighting in the click-click-click of a mouse" (qtd. in Paumgarten 2014). Consider, too, "Groundhog Date," now marketed by Match.com. Participants can e-mail photographs of their ex-girlfriends or ex-boyfriends, so that facial recognition software can find the most similar faces among millions to deliver singles "their type" (Pasquale 2015, 39).[4]

Mirror algorithms know our tomorrows before we do. Patterns of "likes" on Facebook are correlated to the likelihood of whether mothers may malnourish their children (van Dijck 2014, 204). In Will Oremus's "Google Knows What You're Doing Tomorrow" (2013), he argues that new Google products would be able to search your own personal computer, calendar, and websites, photographs across five categories, "flights, reservations, purchases, plans, and photos." Some of this has been developed on a Google Now app on Android phones as well as a "Gmail search" trial that searches users' computers. There is also a slight tone of concern and anxiety here: "Still, there's a good chance that some users will be surprised, as they try out the new features, to see just how much Google knows about them—and how good it's getting at applying that knowledge" (Oremus 2013).

Algorithms are also increasingly used to predict health and medical outcomes, with jacked-up digital biopower systems with gadgets like an app than can predict with two years' anticipation if your child will become a binge drinker "with 70 percent accuracy" (Sampson 2014). That app, of course, relies on compiling a series of risk factors, taken from a questionnaire of "40 different behavioral, biological, and environmental traits." It is another example of how artificial intelligence locates human intelligence, knows about us, can predict our behavior, and becomes a supple tool for self-improvement. Such computational processes also undermine a sense of agency in determining social relationships because we lose the sense of enchantment, even if that is an illusion, that friendships and love relationships are built on spontaneous human feelings, desires, and attraction. The *New York Times* reports various roommate pairing programs that universities use in the hopes of avoiding the most expensive issue, which would be one of the two students dropping out due to roommate issues (Singer 2014).

The quantification of personalized prediction is not new, but its systematic mechanization is; as theorist Adrian Mackenzie (2015) puts it, "The production of prediction is not automatic, although it is being automated" (444). The capacity of data mining allows for a hyperpersonalization of information; put differently, the speed, intensity, and extraordinary mass of available data make algorithmic knowledge feel immediate and wondrously curated. Moreover, self-tailored prediction makes certain kinds of random data into stable classifications in order to forecast new knowledge. Mackenzie suggests, "The classifications may be rather arbitrary or highly artificial (for instance, the group of people who own dogs and click on Honda ads while Wimbledon is on) but they must be relatively stable." (441). In other words, we find that these arbitrary classifications are subsumed by the vectorization of data that produces other meaningful, though also sometimes arbitrary, relationships. The arbitrary nature of this data collection means that users on various websites might also be less skeptical about how seemingly meaningless data are not only collected but also meaningfully correlated to other data. This "emplotment of prediction," in turn, requires a living and constantly changing network of labor (analytical work) and information that will produce and recombine to generate new predictive relations (Mackenzie 2015, 443). While the infrastructure may result, perhaps quite hopefully, "in the diversity of social production or inform new collectives" (Mackenzie 2015, 443), it seems to undermine the likelihood that subjects will recognize how minor actions (clicking on an ad for Honda) might shape how they are targeted not just as consumers but as voters, as potential criminals, as at-risk health conditions. Such forms of predictions solidify the filter bubble because they extend it into the future.

In addition to stabilizing only certain classifications, algorithmic prediction has become a way of managing the future, as it is folded into predictive governing and the patrolling of citizens, such as no-fly lists and criminal recidivism algorithms (McQuillan 2015, 567–568). McQuillan argues that the paradox of algorithmic governance is that the objective is to limit, not expand, the choices for citizens, and thus algorithms represent a political technology of limitation and curtailment, though they are seen as widening opportunities for engagement, consumption, and experience. Moreover, in March 2018, Facebook was in the midst of scandals as Cambridge Analytica mined over fifty million profiles' user data in order to pinpoint how and to whom to market Donald Trump as a candidate, and Congress questioned Facebook CEO Mark Zuckerberg for selling user data and for allowing unseen levels of fake news on the platform. In the case of Cambridge Analytica, an apparent personality quiz was used to correlate traits to political preferences, and, more crucially, "scrape some private information from their profiles and those of their friends" (Rosenberg, Confessore, and Cadwalladr

2018). While Rossiter and Zehle (2014) did not predict that algorithmic centers like WikiLeaks or Anonymous would produce a "raid on modern institutions of control (the state, firm, military, union, etc.)," they did argue that the new sites of vulnerability would be "the algorithmic architecture that increasingly determines the experience and conditions of labour and life" (127). Yet the election scandals with Facebook reveal that political elections—the very epicenter of modern institutions—now manipulate algorithmic data from social media.

Derek Thompson (2014) tackles the issue in *The Atlantic* by focusing on how Facebook and Amazon use algorithms which are defined as "a piece of code that solves a problem." For example, Facebook might use a combination of what users have responded to in addition to some demographic information about the user in order to cater the News Feed to that user. Amazon developed its own special algorithm called "item-to-item collaborative filtering" in the 1990s: "Amazon built an index of items that customers tend to purchase together." It's more based on their inventory than its users. Still, Thomas offers an example of a disgruntled Amazon user whose "Recommended" feed was 100 percent based on a book he had purchased: every single recommendation was by that same author. Thomas puts it this way: "It's the recommendation of a software index that has no idea who you are." This is precisely the point: the rise of the algorithm economy means simultaneously creating the desire for it, which is what this user frustration seems to imply. Consumers, who are increasingly and constantly encouraged to see everything as tailored and self-making in our popular culture of voyeurism and narcissism, want to feel special and unique as they browse online. It is not about the proper algorithm knowing which books to recommend. It is about the proper mirrored algorithm validating the illusion of endless individuality while actually synchronizing consumer needs, wants, and affect. Thomas asks us to "embrace a new version of intimacy" where "the machines have to know us." Yet this means succumbing to the very parameters that feed all the way up to the surveillance state (Vannini 2015).

The issue here is a locus and a uniquely human structural deficiency: our minds cannot necessarily process the required manipulation of day-to-day tasks that modern life demands. Or so we will be increasingly told. But what this also means is that the locus of knowledge shifts to software. Now certainly a computer model replaced forms of human cognition over the past fifty years in ways that are unthreatening: nobody was questioning that

NASA calculations needed specialized programs or that finance moved toward crunching billions of data points on the market. What is new here, however, is the banality of it, the everyday habit of thinking that is shifted to the digital and electronic. The article also jokes that we are "a long way" from asking Google, "What *should* I do tomorrow?" (Oremus 2013). Yet what is evident is that we increasingly rely on technology, on machines, for simplifying our lives but not as our moral understanding of the world and for forecasting increasingly personal habits and customs. But there is far more at stake. Nick Couldry (2017) warns the "infrastructure of surveilliance is in tension [. . .] with the notion of autonomy that provides the reference point for most visions of democractic life" (183). To Couldry (2017), then, algorithm society threatens democracy precisely because it undermines "the voices of individuals, and their accounts of the lives they have lived as autonomous individuals" (187; see also Couldry and Mejias 2019).

This massive paradigmatic shifts our modes of everyday epistemology toward the quantifiable and countable, and thus, in turn, commodifiable. As media scholars Paolo Totaro and Domenico Ninno put it, "The logic of numeric functions enters the practical world, often unseen, and firmly takes root in everyday life and our consciousness" (Totaro and Ninno 2014, 30). Modern forms of rationality employ "the mathematical datum as an instrument to communicate a logically coherent interpretation of sensible experience" (Totaro and Ninno 2014, 46). Both algorithms and practical forms of calculation represent a "mathematical datum" that becomes part of everyday life. They suggest that this modern inclination toward "formalization" accelerated the incorporation of algorithms into production. In fact, they suggest that algorithms' centrality was more a result of a push toward formalization than a result of a push toward mechanization, also hastened by a bureaucratic infrastructure and commodification of calculation (32). And, of course, algorithms have radically shifted and remade capitalist economic structures, both production and consumption regimes (Totaro and Ninno 2014, 44). Italian theorist Stefano Harney (2014), in a collected edition called *The Algorithms of Capital: Accelerationism, Consciousness Machines and the Autonomy of the Commune*, likens the algorithm to Marx's "automatic subject" ascribed to the commodity as part of exchange, and says, "That magic, that lightning, takes the form of the algorithm, and the algorithm encourages capital to follow this mortal fantasy as never before" (114). The agency of certain financial actions, he argues, lies in the algorithms not in bankers or human decision makers alone, but in a deep entanglement of machine and human knowledge (117). In this, Harney (2014) helps us make sense of our late capitalist regime to its epistemological underpinnings: the way in which mirrored algorithms mystify information. But before we unpack why algorithms are mystified, we must first see why the customization process traps people in endless loops of their own making. Put differently, in order for the algorithm to become otherworldly, it must first be inner-worldly.

Filter Bubbles

Until now, we have covered how mirror algorithms track the individual and mine for data that can be commodified and customized to the user, for the present or the future. But what happens when these highly tracked users seek information about the world? Google represents another foundational mirror algorithm in which the user's own worldview might already be embedded in search results, so much so that the results already reflect the user's own political beliefs, based on past searches, and friends, based on friends in the user's social networks (Bozdag 2013, 219; Bozdag 2015; Lupton 2015; Pariser 2011; van Dijk and Hacker 2018). While most users may know that their search history and popularity inform the algorithm, the Google algorithm may also embed the user's zip code, IP address, social media likes, consumer purchases, ad-responses and possibly demographic information on age, gender, or race (West 2017). Plus, Google technicians may override the algorithm in order to either blacklist or whitelist certain sites (Bozdag 2013, 217). "Filter bubble" is a term for this kind of biased results in which user results show only those results that already align with the user's worldview. Eli Pariser's *The Filter Bubble* (2011) examines the political fallout of citizens who were socialized into siloed spheres of self-referential information or "identity loops" (127). Pariser has been credited with predicting the 2016 election of Donald Trump and warning that these online echo chambers could affect democracy (Jackson 2017).

As a whole, Google is rather secretive about the inner workings of its algorithm, so such details are largely unknown (West 2017). This has partially been the result of an ongoing race to either hack or outsmart Google searches, thus rendering Google's algorithm "a behemoth whose inner working are a near-complete mystery outside the company" (Hancock, Metaxa-Kakavouli, and Park 2018). Koopman (2015) reminds us of the metaphoric watchtower, drawing from the Foucauldian panopticon, yet also suggests it is a problematic metaphor because it is a symbol of visibility, whereas most algorithmic tracking functions are invisible: "It is not so much as the panopticon as pan-analytic." Datafication relies on people's trust in institutions or at least their preference for convenience and service with the sacrifice of personal data.

Engin Bozdag (2013) analyzes how Facebook and Google, among others, personalized their sites with "algorithms that tailor information based on what the user needs, wants and who he knows on the social web" (209). What the user sees on Facebook is dependent on an algorithm that considers interactions among friends, comments, likes, and kinds of posts. By extension, if users are interacting with a particular political or ideological group, as in liking posts or clicking on media stories in their newsfeed, the algorithm will produce more of the same. Theorizing that "information mediators" like Google and Facebook "control the diffusion of information for millions of people," Bozdag (2013) theorizes: "The gatekeeper controls whether information passes through the channel and what its final outcome is, which in turn determines the way we define our lives and the world around us,

Figure C.1. Image of Luigi di Maio, 2019. Reprinted with permission of Riccardo Antimiani.

affecting the social reality of every person" (211). The users have less access to sites with alternative or contradicting views and "spread falsehoods, and promote polarization and fragmentation" (Sunstein 2017, 5). It will keep the user within an information network *already* aligned with the user's view of the world and hide or "silently filter out" opposing political views (Bozdag 2013, 218).[5]

Filter bubbles overpower the user's desire to try to move outside the user's information circles because algorithms are working implicitly. The user, then, sits in a kind of silo. Bozdag's (2013) term "cyberbalkanization" is a helpful analytic for the reduction of the citizen's isolation (221), as is Sunstein's (2017) notion of "echo chambers" (6) and Pariser's (2011) "identity loops" (127). Because they expand political polarization, narcissistic worldviews, and siloing citizens from alternative views, filter bubbles enable disinformation. Consider the Facebook users whose feed is already only visualizing the news based on their most-liked friends and existing news stories. A new story—whether about Giuliani's earthquake prediction or Hillary Clinton's emails—would be more likely to appear before their eyes. Their click only further aligns them with more fake news. It's like receiving robo-calls: when you answer the phone, the robo-calls multiply.

The mirrored window society aims to show it is not only isolation, but a self-replicating mirage: the mirage of obtaining new information when the algorithms are filtering through the existing individualized tastes, likes, and views, always—yet unknowingly—reflecting users back to themselves. The feedback loops indirectly validate the user's existing opinions and beliefs because it falsely appears as if information is aligned with the user's views and preferences. Plus, users just look prettier, thanks to enhanced selfie algorithms. Like the mirrored altar in *Io C'é*, mirror algorithms are created out of sight, making their glossy sheen a new site for misunderstanding the underlying structure of information production.

The Mystification of Algorithms

In "The Cathedral of Computation," media studies scholar Ian Bogost (2015) asserts that algorithmic culture "is not a material phenomenon so much as a devotional one, a supplication made to the computers people have allowed to replace gods in their minds."[6] *Algorithmic Culture* author Ted Striphas (2015) has called some of the biggest algorithmic corporations—Amazon, Google, Facebook, Twitter, and Netflix—"the new apostles of culture" (407).[7] Several scholars have observed this increasingly common enchantment of information systems, such as "algorithmic gods" (O'Neil 2016, 8), the "idolatry of data" (Weseltier 2015), "tales of autonomous algorithms" (Seaver 2018, 379), "the godlike quality of some roboprocesses" (Gusterson 2019, 39), and the "magic of code" (Finn 2017, 51). But how exactly does this form of intelligence become godly? In order to understand why data customization and filter bubbles have such high potential for remaking politics, we must first understand the systematic and persistent mystification of algorithms. First, computational systems regularly rely on hiding the humans involved in the design and process, and second, they hinge on the appearance of science as objective and "disinterested" (Rossiter and Zehle 2014, 118).[8] The apparent "disinterest" of algorithms is also tied to their ability to appear as objective forms of computational knowledge, which strips and masks the role of human agency in their making and ultimately endows the nonhuman with an enchanted power over human agents.

Dehumaning the algorithms comes first, followed by an impoverished public understanding of *how* they work. Exploring the production of a music program algorithm, Nick Seaver (2018) argues that while human programmers and intelligence are very much involved in algorithmic programs, they often seem invisible to users.[9] The system and the humans in the system become obscured. To Bogost (2015), computer science that supposedly promotes skepticism and rationality actually produces "a new type of theology" in which algorithms become false idols because hardly anybody knows what lies "under the hood." We find a familiar logic here, an insight borrowed from classic anthropological studies of magic and science: the less that is known about how something works, the greater the likelihood for its mystification and deification (Frazer 2009 [1890]; Malinowski 2015 [1954]).[10]

There are dire consequences to mystification, both micro and macro. Besteman (2019) suggests the "consumer-subject" is positioned in such a way that makes "few opportunities for contestation, appeal, negotiation, or refusal," especially considering the lack of understanding of how these processes work and due to the lack of legal precedent for accountability over data (9, 17–18).[11] Gusterson (2019) analyzes shocking news reports of people who drove their cars directly into bodies of water—oceans and lakes—just because their GPS "said so." The authority of the car navigation system seems to undermine everyday common sense and rationality in spectacular ways (27). Just as the words of scientists undermined Aquilans' sensory experience and familial narratives on earthquake danger, so too does the GPS voice seem to anesthetize individual judgment to environmental hazards. Both instances of car travel and impending earthquakes obviously pose fatal danger to citizens, and both can be read as precautionary narratives: as the center of knowledge shifts to "data" and that knowledge appears as godly, it also undermines individual judgment of and sensory perception of risk and danger. Such self-endangering choices also rely on the faith everyday citizens endow in these disembodied technologies and scientific narratives.

In this sense, we might not see these forms of enchanted science as particularly new but rather as a continuity of how social actors make sense of phenomena from an underinformed position. Latour theorizes this as a false elegy for enchantment in *We Have Never Been Modern* (1993), which argues that the notion of secular, disenchanted, faithless science is fundamentally a Western myth that assisted in the crafting of modernity, which relies on a separation of the human and nonhuman. Similarly, in *The Faith of the Faithless: Experiments in Political Theology*, Simon Critchley (2012) likens political history to an ebb and flow of faith, "a series of *metamorphoses of sacralization*" (Critchley 2012, 10). For Critchley, politics demands a literary analysis of faith, not as a form of divine belief or metaphysical conviction but as an "enactment of the self in relation to an infinite demand" (13). The paradox, then, is that the enactment of the self does not rely on the rational and calculating subject, but rather subjects must be compelled by both rationality and faith (Critchley 2012, 19). The algorithm, which undergirds the political and scientific apparatus, thus aligns itself with this appeal to both rationality and faith, a belief in its power that goes beyond reason.

While the mystification of the algorithm may intensify as belief stands in for technical knowledge, another coproducer of mystification might be the disembodied experience of neoliberal capitalism. The process hides not only the data customization and underlying mechanisms but also the accumulation of wealth into fewer hands and the expansion of inequality. Gusterson (2019) suggests that these new beneficiaries, these new centers of capital and power, are often invisible to citizens (32).[12] We might argue, then, that data capitalism has merely intensified certain trends of neoliberal capitalism—"occult capitalism" or the "zombie economy"—with its unpredictable rise and vanishing of wealth and unseen financial processes (Comaroff and Comaroff

2000; Giroux 2010).[13] Mystification of knowledge is also spurred as algorithm culture intensifies the value of the immaterial and commodified body: the immaterialization of day-to-day communication as inversely proportional to the affective and sensory experience of others' bodies.[14] It is the acute realness of email, the Internet, YouTube videos, reality television, film, and twenty-four-hour news that more than any other time gives us a real experience that is more lucid (large, HD televisions and screens) and seems closer than ever to three-dimensional corporeality but remains two-dimensional, elusive. It is also the cultural investment and massive amounts of psychic labor that ask modern subjects to invest, affectively and cognitively, in these encounters and their production of the real that produces a dissonance between the corporeal and the immaterial.

The Clown and the Android

What kind of new political forms emerge from mirror algorithms, a world in which information is customized yet systematically hidden and, in turn, mystifying the very algorithmic functions that control citizens? Solla's (2011) work on "ubu-esque" power describes Berlusconi's absurd yet potent political style, adopted as Alfred Jarry's titular king from his 1896 play, *Ubu Roi*. The figure of King Ubu has already served as an analytic for forms of grotesque normalizing power for Foucault. It is worthwhile to reconstruct the deeper genealogy of *Ubu Roi* particularly in relation to forms of governance and institutional power. For Foucault (2003), ubu-esque power was part of biopower, which relied on processes of normalizing large populations, which also entailed the surveillance of medicalized and deviant bodies. For Foucault, the psychiatrist represented this form of ubu-esque or grotesque and "defensive" power to screen bodies as potential disruptions to social power, especially in their use of "obscure and ridiculous terminology: diagnoses of conditions such as *alcibidadisme, bovarysme, don juanism*" (Islekel 2016, 185). While such "odd discourses" were humorous and absurd, they were nevertheless powerful and could control life and death of subjects. The extension, then, in terms of ubu-esque power for political leaders, also bears this dualism between recognizably absurd and yet having life-or-death consequences. For Foucault, ubu-esque power was embodied by experts and holders of knowledge, not the sovereign or top-down rulers like Ubu Roi, the king in

Jarry's original play. Yet, following scholars like Solla (2011) and Lukes (2016), my own understanding of ubu-esque power must contend with why political leaders, on a national scale, have come to be increasingly absurd in late modernity. Thus the tension between the ridiculous and the governing is central to ubu-esque power whether it is embodied in a medical figure or the national state leader.

Frank Kenesson (1993) adopts Jarry's king as a lens to rethink European nationalism. Of Jarry's leader, Ubu Roi, he says, "As easily as Alfred Jarry made of the King of Poland, Ubu Roi, a paradox of fevered inaction, exalted fecklessness and eternal precariousness, as well as minister of the violent caprice they breed; that easily is condemned to represent passivity, complacency and prejudice . . . the self-delusion, melancholy and methodical cruelty of his nation's failed political experience or theatre" (436). Kenesson connected the nationalist movements of the late 1980s and 1990s: "It unites and fortifies a people in uncertain moments of confused transition . . . precisely because each member sees it somewhat differently. . . . It is, to each, the very image of himself" (436). For Kenesson, nationalism harvested egotistic self-interest as a kind of collective illusion, as uncertainty allowed for members to reshape the group as an internal mirage. H. N. Lukes (2016) refers to American president George W. Bush as the epitome of ubu-esque power (331) and shapes a new body politic "who would rather ego-cast than vote" (331). The psychoanalytic lens offers a similar conceptual idea as cultural theorists of technology, that is, the way in which the citizen-subjects become trapped in a chain of reproducing the self, or worse, of projecting their prismatic worldviews on the ubu-esque leader.

As a political process, then, delusional individualism can be marshaled toward nationalism, which may be why, in a political age with several competing forms of populism across Western nations, we once again see the ubu-esque leader. Building on some of these logics, I am suggesting algorithm society produces more absurdist governance, with a strong nationalist undercurrent, because it allows for even more entrenched replication of self-interest and individually catered knowledge. Citizens garner seemingly neutral information about the world while unaware of how that information has been customized, commodified, and tracked. In the world of Big Data and algorithmic culture, the insular individual replicates and reproduces the individual's world yet is even less aware than Kenesson's citizen-subject.

Today's user-citizen is unaware precisely because the mechanisms of digital knowledge and algorithmic technologies hide the way the individual's history and preferences are repackaged as if they were objective functions of technology, like built-in enhanced selfies on a camera phone, or as if they were functions of a group, not an individual, as in Google searches or Uber. The future, then, may find us with increasingly narcissistic and ineffective leaders because knowledge production is bound up with mirror algorithms, reproducing a preexisting political ideology in ways that are ever more hidden and unrecognizable.

Let us return, at last, to reconsider the theory that Five Star Movement leader and Deputy Prime Minister Luigi di Maio was perceived as an algorithmic creation for one of the world's first political parties created in and through digital technologies. In addition to showing why algorithms were essential to the rise of the Five Star Movement, let us reexamine the theory that Di Maio was not a leader who happened to rise in rank in the Five Star party but rather a "populist android" (Minuz 2018). Further, tech writer Minuz claims,

> Di Maio is like Frankenstein, made from the Italian leftovers of social movements and civil society, from coins to Raphael's *girotondi*, from violent population to the people of the fax machine, from pitchforks and from Milan swimming pools, from territories, from drills, from veganism, from common goods, from a most formidable accumulation of anti-scientific, conspiracy and anti-economic theories, from *Iene* (The hyenas), from legal encounters, from anti-Uber taxi drivers, from La7 talk shows, from the happiness decline and from the still enthusiastic destruction of the GDP, anti-Vax, anti Tav, no Global, no Bilderberg, no Bolkestein, no euro, yes we can, let's have a referendum on Facebook. (Minuz 2018)[15]

Minuz's Frankenstein Di Maio is a conglomeration of Italian classics and millennial movements: the Renaissance masters like Raphael but also the popular television programs. The theory about Di Maio mimics a kind of computer output of social movements and conspiracy theories in Italy culminating in what Minuz calls "the incontestable virtue of the populace." Like Frankenstein's patchwork body of reused parts, so, too, are the conglomeration of political stances his appeal incorporates: elitist (Milan swimming pools, drills) and antielitist (no Bilberberg, no Global, anti-Uber), tech savvy (Facebook referendum) and tech phobic (fax machine), urban and rural, and high and low education.

The algorithmic Frankenstein also poses a sharp contrast to Berlusconi's coming-of-age political form: his 2001 self-published biographical book, *An Italian Story* (*Una storia italiana*), which was distributed by mail to homes across all of Italy (Concina and Costa 2001). Berlusconi paid many millions to make sure his glossy and 125-page photo-filled life story would end up in the hands of Italians. The book was adorned with a photo-collage cover and many personal photographs, and citizens could read sections in it like "Character and Passions: The Life of Silvio Berlusconi, Childhood, Adolescence, and Schoolmates" ("Il carattere e le passioni: La vita di Silvio Berlusconi, l'infanzia, l'adolescenza, i compagni di scuola"); "Lifestyle: How He Dresses and What the Leader of Forza Italia Loves" ("Lo stile di vita: Come si veste e cosa ama il leader di Forza Italia") and "Constructing an Empire" ("Costruire un impero"). In *An Italian Story*, we see the material strategies of the then right-wing populism as visual and narrative. The book employed strategies from television, which was surely how most Italians at the time encountered the leader. Rather than list editors or writers, the book credits "artistic directors."

Still, the book was animated by populist principles of this late 1990s and early 2000s period such as Berlusconi's status as political outsider, his individualism, and perhaps most of all, the value of his self-made wealth and fortune, as in the very first line: "From his father Luigi, all together Milanese, an ancient kind, Silvio gets his sense of duty, his love of work, his capacity to sacrifice, his respect for giving one's word." The book ends with a letter to citizens from Berlusconi in which he addresses the upcoming elections and his vision of "a free Country, prosperous and fair where doors are open to hope for everyone" (*un Paese libero, prospero e giusto dove per tutti sia possible tenere la porta alla Speranza*). This final letter was the utmost gesture of personalization, like a national address to all citizens, but materialized in print form to convey a more spectacular sense of his wealth and reach. Indeed, the very emergence of his book also tells us a story of how political leaders were made as a function of how information about ideal leaders was culled and redistributed to the population. Berlusconi's book writers knew what would work based on not just voting patterns for *Forza Italia* and polls but also his company Mediaset's viewing patterns and television ratings, with television the most consumed media at the time. The style of narrative, and its form, the booklet, centered on a single protagonist and repackaged Berlusconi by editing and shaping the words and images to best suit the values he was fashioned to represent.

To be sure, Di Maio is not Berlusconi, even if Berlusconi represented a kind of ubu-esque leader in his clown-like gaffes and outlandish scandals in the 2000s. Moreover, he and Berlusconi diverge not only on their political platforms but also in Di Maio's overall reputation that verges on naiveté which is nearly opposite from "The Knight's" self-serving cunning. But Di Maio still may represent a kind of "fecklessness" and "fevered inaction" of the ubu leader (Kenesson 1993, 436). Di Maio is like Ubu in the technological incompetence in that unlike other forms of algorithmic intelligence in which human labor is hidden, his widely appealing personality is marred because he seems, at least to some, too perfectly engineered. Whereas the algorithms I have discussed earlier control and shift power toward corporations and the wealthy minority because the human labor is hidden, Di Maio's perfect likability betrays the most fundamental rule of algorithmic mysticism: hidden human design. In fact, in this narrative, Di Maio appears as if Casaleggio, the creator of the Five Star digital platform Rousseau, had deliberately mined user preferences in order to design the perfect populist candidate, one who could be likable from all divergent populist corners: the antiscience, the talk show set, even the deep global financial conspiracy crew ("no Bilderberg"). Di Maio's "on the nose" design renders him a fabrication of the algorithm age, a Frankenstein of data mining.

Perhaps the new Ubu ruler for the age of mirrored windows is the "populist android." Just as the age of politically manipulated televised media gave rise to Berlusconi and the figure of the clown, so too, then, the age of the politically manipulated Internet christens the android. The clown and the android are twin ubu-esque figures separated by political ideologies and divergent information technologies showing us, once again, that how people learn about the world shapes who they think should rule it. Both the clown and the android tell stories of falsity, and a world in which truth must be scrutinized and doubted, and both figures are revealing of their unique and hegemonic form of information gathering. The clown conjures the visual image of the painted and masked face which plays on the televised camouflage and visual trickery necessary for Berlusconi. The android conjures the posthuman algorithmic technologies of Five Star and Di Maio. And like Massimo and his sect of "selfism," Di Maio might even tell his own followers directly that he was chosen because of their will, their preferences as revealed on the Rousseau platform (*Io C'é* 2018). And like Massimo's followers who denied this truth and avowed him as prophet, the majority of Five Star followers would likely scoff at the android claim and avow Di Maio as an incredibly authentic representative. In this sense, Di Maio gives us a glimpse of a new genre of algorithmic populism we may see increasingly populating political life and animating political leaders in liberal democracies.

Notes

Introduction

1. Berlusconi served as president of the Council of Ministers of the Italian Republic (presidente del Consiglio dei ministry della Repubblica Italiana), a title that is often shortened to "president of the council" (*presidente del consiglio*) or referred to in English, which I will adopt as "prime minister of Italy."

2. A *rettore* in Italy is most similar to a university chancellor or president in the United States.

3. In the American context, Colbert's truthiness was a response to the political disenchantment with President George W. Bush and the Iraq war. Colbert elaborated on his definition of "truthiness" for an interview with *AV Club*: "It used to be, everyone was entitled to their own opinion, but not their own facts. But that's not the case anymore. Facts matter not at all. Perception is everything. It's certainty. . . . I feel a dichotomy in the American populace. What is important? What you want to be true, or what *is* true?" (Rabin 2006). Colbert picked up on the way that the manipulation of truth becomes enmeshed not just in politics but in a sense of entitlement. While we might think that a certain truth of the reality of what happened represented a fundamental limit for culture and political institutions, we find that is not the case. The 2016 American presidential

election and the candidacy of Donald Trump ushered in a new kind of sharper, more blatant form of truth-denialism. Trump was already known for leading the American "birther" movement, which propagated the false idea of Barack Obama's fraudulent American citizenship. During the elections, Trump's lies accumulated from denying his support of the Iraq War to defining his relationship with Russian president Vladimir Putin. Indeed, the *New York Times* published a full-page roster (December 14, 2017) of Trump's lies since his presidency began, including lying about the size of crowds attending his inauguration. In one notable incident, his press secretary Kellyanne Conway defended the press secretary statements on the inaugural crowds as "alternative facts." In April 2017, Americans participated in a "March for Science," with the explicit defense of a shared, objective reality, including signs that stated, "Science is not an alternative fact" and "Science is real" (Chow 2017).

4. Hetherington (2017) problematizes the genealogy of "post-Truth" by calling out assumptions that information, resistance of state tyranny, and growth were inherently and fundamentally mutually reinforcing. He exposes the contradictions in American post-Trump explanatory narratives: "In the American context, Post-Truth is really a story about the collapse of a geographic firewall between reason and unreason that liberals have held dear since the beginning of colonialism."

5. He also argues that these people then verify truths through a social means by "seeking recognition through the technical apparatus, the web, that allows for the expression of individuals rendering them irrelevant (precisely because one is just as valuable as another) and instead becoming the microphysics of power, funneling all those ideas—different, but all equivalent—into a gigantic 'I like,' an apparatus that counts for the numbers it expresses and that, together, shows a totally atomized public opinion."

6. In agreement with Marshall et al. (2015), Shaheed Mohammed (2012) resists what he dubs modern narratives of the information age and argues that disinformation produces ignorance, which, he argues, becomes folded into notions of "'tradition,' 'heritage,' 'religion,' and 'culture'" (14). Further, the rise of anti-intellectualism in the West increases skepticism toward the scientist: "The scientist or researcher who generates, interprets and applies scientific data is not necessarily more valued than the paparazzi who generate information in the form of scandalous photos of celebrities" (24). While I would suggest the value of celebrity information may be especially acute in the United States, I do think a devaluation of the scientist is as relevant and important in the Italian context. In my examination of the trial in L'Aquila, I attempt to tease out the tension between trust and skepticism toward scientists as information experts.

7. The Comaroffs (2000) agree that political theatrics and "magicality" have become intensified, more ritualized: "[Ruling regimes] have become caught up in cycles of ritual excess in which ceremonial enactments of hyphen-nation, alike in electronic space and real time, stand as alibis for realpolitik—which recedes even further as its surfaces are visible primarily through the glassy essence of television, the tidal swirl of radio waves, the fine print of the press" (329). Paul Kennedy's (2017) notion of "vampire capitalism" also draws on an occult metaphor to describe how wealth is not reinvested into markets, which drives the accumulation of capital into the hands of fewer elite citizens, "the global plutocrats" (42). David Graeber (2012) notes a "qualitative change in the magnitude of political dishonesty" in American politics, which he links to the embeddedness of the

Republican Right in Evangelical Christianity: "a theological version of an essentially magical style of political performance." Graeber refers to a famous article on Bush by Ron Suskind, who examines George Bush's presidency as a "faith-based presidency" in which "open dialogue, based on facts, is not seen as something of inherent value." Suskind was told by one of Bush's aides that liberals were a "reality-based community" and countered that the American "empire" could actually amend this naïve assumption, telling him, "We're an empire now, and when we act, we create our own reality." Suskind's reporting that Republicans saw reality as nothing more than a pliable mass from which political performance would be best unhinged was, according to Graeber, confirmed by liberals' beliefs about the Bush presidency. Yet this puts liberals in a bit of a predicament because it means they are siding with the "tradition of Enlightenment empiricism" when they have simultaneously dedicated themselves to "trashing the very idea of objective reality." Yet as much as Graeber shames liberals for their Foucault-worship and partial funding by celebrity culture, trying to equate the Left's take on reality production with the Right, he misses the extravagant and bold invention of antiempiricist propositions that sustain contemporary right-wing politics. To critique the singularity of truth and pay attention to its production does not mean to throw away some grounding in empirical truth whatsoever, nor does it mean we, as Graeber puts it, merely hold onto "the notion that performance really is everything." Still, Graeber ultimately sides with the view that the power of political performance can be historicized, a fascination that for him comes out of the American 2000s "bubble economy," like for the Comaroffs, which derives from neoliberal economic regimes.

8. Marcus quotes from Michael Serres's treatment of "the Parasite" in which Serres offers the following:

> To play the position or to play the location is to dominate the relation. It is to have a relation only with the relation itself. . . . It has relations, as they say, and makes a system of them. It is always mediate and never immediate. It has a relation to the relation, a tie to the tie. . . . The whole question of the system now is to analyze what a point, a being, and a station are. They are crossed by a network of relations: they are crossroads, interchanges, sorters. But is that not analysis itself: saying that this thing is at the intersection of several series. From then on, the thing is nothing else but a center of relations, crossroads, or passages. It is nothing but a position or situation. And the parasite has won. (8)

9. Shapin (1994) also recognizes the significance of distrust: "Each act of distrust would be predicated upon an overall framework of trust, and, indeed, all distrust presupposes a system of takings-for-granted which makes *this instance* of distrust possible. Distrust is something which takes place on the *margins* of trusting systems" (19).

10. Giovanni Orsina (2014) argues that an inherent part of Berlusconi's success was reflecting positivity about Italian national identity and society: "He made an ideology out of the positive character and self-sufficiency of civil society and turned it into a propaganda and consensus-building tool" (62). Moreover, he describes him as a "skeptical politician in a historical context marked by the excess of politics of faith" (68). Put differently, Orsini recognizes that Berlusconi's success was based on a kind of doubtful

but persistent capacity to win political consensus without resorting to political tactics around the communism and postcommunism poles.

11. Berlusconi-owned media are known for their systematic objectification of women as ancillary, beautiful, sexualized, vapid bodies that adorn the screen, which amounts to "the hegemony of the culture of beauty in the representation of women . . . and in the sexualization of culture, saturating the media landscape with a mainstreamed sexual and pornographic discourse" (Benini 2013, 89). The political regime of Berlusconi is thus a profoundly patriarchal one. Its mediatization relied just as much on the gendered subjection of women as it did his own control of news information and dissemination of narratives advantageous to his political objectives.

12. Furthermore, I take a cue from previous studies that show how Berlusconi's control of media and public discourse actually gave rise to some surprising new forms of media activism (Berardi and Jacquemet 2009). Franco Berardi and Marco Jacquemet (2009) show how independent radio, small and illegal micro-TV stations, and Internet activism are against-the-grain forms of media that grew out of the homogenization of Berlusconi-controlled media.

13. Since 2017, comparisons between Donald Trump and Berlusconi have been common in public media (Jebreal 2015; Kirchgaessner 2016; Clementi, Haglund, and Locatelli 2017; Doyle 2017; Newman 2017; Taylor 2016). Berlusconi, like Trump, has been famous for his extreme wealth and luxurious lifestyle built on "mafia-linked real-estate developments" (Newman 2017), gaffes and politically offensive humor, antagonism with the press, misogyny, corruption and unethical bending of laws, and governance for self-enrichment and to avoid incrimination (Orsina 2014; Severgnini 2016). Tapping into frustrations with established politicians, both men built a base on the notion that their business acumen makes them superior to insider politicians, despite their repeat bankruptcies and failures (Taylor 2015; Cohen 2016). Both men, though seemingly embodying non-Christian values, were supported by conservative religious groups to forward their own agenda: Berlusconi by the Vatican and conservative Catholics, Trump by white evangelicals. Berlusconi, like Trump, baffled the Left, and the Italian Left struggled to understand his popularity among people whom his policies would most disenfranchise (Kirchgaessner 2016; Agnew and Shin 2016). Italian historian Paul Ginsborg remarked that both Trump and Berlusconi use patrimony as a political strategy, as they both had "the need to defend themselves through changing the hands of power within democracy and making American democracy and Italian democracy autocratic" (Taylor 2017). Perhaps the most significant difference, however, is that while Berlusconi was the owner of television media, Trump's manipulation of media was always secondary, as the darling of Fox News but not its owner, and this distinction matters. Some of Trump's lines seem taken directly from Berlusconi's script: in July 2017, in the days around a G8 summit in Hamburg, Germany, Trump tweeted, "The WEST WILL NEVER BE BROKEN. Our values will PREVAIL. Our people will THRIVE and our civilization will TRIUMPH." In 2001, also around a G8 summit, that time in Genoa, Italy, Berlusconi was criticized when he said, "Western civilization will conquer all people" and "We must be aware of the superiority and force of our civilization" ("Berlusconi, 'Occidente, civilta superiore'" 2001). Trump's battles against the American press seem tame considering Berlusconi owned his own media conglomerate, and censored and fired particular journal-

ists who spoke against him (Kirchgaessner 2016; see chapter 1). Trump's employment of his daughter Ivanka and son-in-law, Jared Kushner, are derivative considering Berlusconi sidestepped similar conflicts of interest by letting his family run his company while he was in office (Newman 2017). Berlusconi's downfall came after serving nine years as prime minister between 1994 and 2011, and being convicted for soliciting sex from an underage prostitute and tax evasion. How Trump exits American politics remains to be seen. Already trying to one-up Trump on his sexist humor, Berlusconi has commented that the best thing about Trump is his wife, Melania (Balmer 2017).

14. In line with studies on mediatization and political participation (Dahlgren and Alvares 2013; Beck 2015), I also reveal in chapter 1 what conditions in Italy made the success of Berlusconi more likely in the late twentieth century. The short answer is that a combination of historically theatrical politics and Italy's long-cynical and politically weary population, not to mention a much more rapid pace of neoliberalization, primed Italy more than other western democracies for a populist truth-bending clown leader.

1. Manifest Disguise and Mediatized Politics

1. This chapter is derived, in part, from an article published in *American Ethnologist* in 2013, available at https://doi:10.1111/amet12021.

2. France's Jean-Marie Le Pen exemplifies this notion of traditional right-wing populism (Fieschi and Heywood 2004, 292).

3. Berlusconi was given the title "Knight of Work" (*Cavaliere del lavoro*) in 1977 by President Giovanni Leone, which has since stuck in public usage.

4. Berlusconi's entrance into politics coincided with the demise of the Communist Party, and he "draws support from a conservative and anti-Communist subculture" (Edwards 2005, 238).

5. In 1991, the PCI split into the Partito Democratico della Sinistra (PDS) and Rifondazione Comunista (RC).

6. The name of the program has been translated as "Creep the News" (Tondo and Leyton 2008), "Strip News" (Bendoris 2007), and "The News Slithers" (Michaels 2007), all of which capture the "fake news" genre (Cosentino 2012). *Striscia*'s emergence also has to be examined in the context of a growing "infotainment" industry (Cosentino 2012, 54).

7. Film scholar Mirco Melanco (1995) cites a 1978 film, *Ecce Bombo*, in which someone shouts, "It serves you right, Alberto Sordi!" implying that it was Italian comedy that had stripped Italians of their ability to resist corruption and political crises. In contrast, Andrea Bini (2011) argues that Sordi represents a shift from comedy focused on the working class to one directed at the indulgence and ineptitude of the middle classes.

8. Chris Smith and Ben Voth make the case that George W. Bush's own self-deprecatory jokes and the widespread humor aimed at him during the 2000 election worked in his favor. Even his ridicule on late-night television and *Saturday Night Live* helped the public "to identify with him better than Gore" and showed him to be a "common man" (2002, 127). This was especially significant given an electorate that was getting

an increasing amount of information from television news parody, late-night, and variety shows.

9. Geoffrey Baym and Jeffrey P. Jones define the genre of news parody as including "many textual forms, from faux news anchors who posture authoritatively at pretend news desks, to puppet shows, sketch comedies, and panel discussions. . . . The programming . . . uses humor to engage with, and offer critiques of, contemporary political life and current events" (2012, 4). I adopt the term "news parody" in this same spirit of comparative political humor.

10. Donatella Campus (2010, 228) analyzes some of Berlusconi's media blitzes, including those surrounding his signing of a "Contract with the Italian People" aboard a cruise ship in 2001 and his forming of his new party, Il popolo della Libertà (PDL), at a Milan rally.

11. Part of Berlusconi's war on the judiciary included accusations that the Clean Hands investigations of the early 1990s were "a long-term strategy of the Communists," a claim that has been much refuted by historians (Edwards 2005, 237).

12. *Striscia* was originally produced on Italia-1, which was owned by Berlusconi's Fininvest, later to become Mediaset.

13. *Odiens* is a play on how the English word "audience" would sound if it were pronounced in Italian. This spelling tactic is a popular strategy for naming shows.

14. Matthew Hibberd (2007) provides a detailed account of media pluralism in Italian broadcasting and how Berlusconi's political career has been shaped by his own role as media mogul.

15. Guzzanti's show attacked Berlusconi and took direct aim at his television empire. The title of her show itself plays on RAI, the acronym of the state-owned national television stations, and the word "riot." Eventually, Guzzanti made a theatrical version of her show and moved forward with her increasingly chafing impressions, disseminating her videos on YouTube, TV talk shows, and other television programs (Watters 2011).

16. The *veline* have become iconic in Italy, incredibly famous and popular among audiences as well as symbols of enduring patriarchal gender norms. They are regularly featured in bathing suits and skimpy costumes, and most nights they begin their dancing on the cohosts' desk. Each year, tens of thousands of Italian women compete to be the next *veline*. The competition is its own show (Ardizzoni 2009). The term *veline* actually refers to directives written on scraps of paper by the Fascist era's Ministry of Popular Culture; today they imply fake news (Benini 2012, 91). Today the term is understood to refer to an attractive television showgirl. While the show reveals political corruption and public scandal, it also simultaneously promotes consumer culture (Cosentino 2012, 55, 64). Moreover, the show promotes the highly gendered commodification of the body and hyperindividualism of late capitalism.

17. The phrase *bunga bunga* refers to a twentieth-century joke about explorers and anthropologists facing "death or *bunga bunga*" at the hands of an African tribe. The punch line reveals *bunga bunga* to be anal rape. Berlusconi revived the term during "Rubygate," his 2010 sex scandal surrounding underage Moroccan Ruby Rubacuore (Karima El Mahroug), who attended his sex parties. During the trial and within popular media, the term referred to a variety of sexual practices: striptease, orgies, and anal sex—but mostly, orgies (Benini 2012, 89).

18. Similar to Gabibbo, Brozo, a "subversive clown" featured on Mexican television for the past twenty years, offers audiences "a bawdy and rambunctious interrogation of politics and current events" (Baym and Jones 2012, 7). One of Italy's top newspapers asserted that Gabibbo represents "Leninist humor," suggesting Ricci was "creating a revolution . . . and stuffing his wallets" (Ricci 1998, 165–166). The comment seemed to stem from an understanding of Leninist humor as duplicitous, though it bears noting that it parallels the *stiob* humor that Dominic Boyer and Alexei Yurchak (2010) explore. But such protests were exceptionally rare when compared with Gabibbo's overwhelming popularity.

2. The Soldiers of Rationality

1. Polidoro (2008), a professor of paranormal psychology (*psicologia dell'insolito*) at the University of Milan-Bicocca, appears frequently on television and radio and publishes widely on paranormal phenomena (Nisbet 2000).

2. Scholarship on science and technology studies has examined global contexts and reimagined how scientific knowledge is constituted and diffused transnationally (Latour and Woolgar 1986; Haraway 1988, 1997; Latour 1987, 1993, 1999; Castells 1996) and, within anthropology, through practices of historical analysis and ethnography (Rabinow 1999; Hayden 2003; Masco 2006; Helmreich 2009; Taussig 2009). Much scholarship examines how the very notion of "belief" emerges as dubious only in terms of the rise of a rationalist and scientific paradigm of knowledge (Hacking 1983; Good 1994).

3. An article in Novella and Bloomberg's organization-based-publication *Skeptical Inquirer* that defined scientific skepticism as not necessarily atheist and affirming that religious belief is not within their primary domain of inquiry ended with a peculiar advertisement: that members remember Committee for the Scientific Investigation of Claims of the Paranormal (CSICOP), which is now known as Committee for Skeptical Inquiry (CSI). and their publication in their wills: "You can take an enduring step to preserve [CSICOP's and the *Skeptical Inquirer*'s] vitality when you provide for the *Skeptical Inquirer* in your will (46)." Though I find Novella and Bloomberg's distinction between rationalism and skepticism helpful, I cannot help but wonder whether their careful and pointed affirmation of religious beliefs was not a way to invite religious subscribers to posthumously donate to their organization.

4. In the eighteenth and nineteenth centuries, rural Catholicism was mixed with various practices of superstition, magic, and paganism (Duggan 1984, 26).

5. One such piece circulating this notion of superstition was David Hume's essay "Of Superstition and Enthusiasm" in David Hume, *The Philosophical Works of David Hume* (Boston: Little, Brown, 1854), vol. iii, essay x, cited in Cameron 2005, 264.

6. They also find that lower and higher levels of "religiosity," which they measure as a variable averaging reported beliefs, practices, and organization memberships, correspond to lower levels of belief in the paranormal. The logic here is that people who are most strictly adherent to Catholic views share fewer paranormal beliefs because the two would become incompatible. Subjects with little religiosity might be more likely to reject all forms of matter not able to be empirically tested and validated, so the rejection of paranormal beliefs corresponds with those having few or no religious views.

7. The practice is also known as *rabdomanzia* with the root words *rabdos* or stick and *manteia* or prediction.

3. The Rise of Algorithm Populism

1. This chapter is derived, in part, from an article published in *Perspectives on Europe* in 2013 (Molé 2013b). The Taylor & Francis journal is now called *Perspectives on European Politics & Society.*

2. Renzi's speech in Italian, translation by author: "Questo atteggiamento di diversita' della posizioni sulla base di qual'é l'onda che cresce. Guarda che é fortissimo nel partito algoritmo. Io seguo cioé che viene in qualche modo spiegato, come se fosse un motore di ricerca sul Google, quello che viene spiegato dai desideri degli utenti. Questo per me non è la mia concezione della democrazia."

3. The Five Star Movement (Movimento Cinque Stelle) had its first big success in 2013: it garnered nearly a third of the national vote in the February 2013 national elections, a moment that for some political analysts was the Five Star Movement's first "mind-boggling" breakthrough (Alcaro and Tocci 2018, 1). Afterward and through the fall of 2013, when new elections were still imminent, Grillo refused to align his party with Italy's leftist party, PD, or Berlusconi's right-wing party, PdL (Il popolo della Libertà). With the strongest electoral performance by a first-time party in Western Europe, in 2013, the Five Star Movement still had a strong national representation, with 109 deputies and fifty-four senators in the Italian Parliament (D'Alimonte 2019, 117). Prime Minister Enrico Letta was appointed in April 2013 after a political deadlock between Grillo's Five Star Movement and right- and left-wing parties; Mario Monti, an economist and technocrat who served from 2011 to 2013, tried unsuccessfully to resolve Italy's debt crisis. By 2018, his party had nearly won the national election, yet no single party took the significant majority.

4. In the March 2018 national election, led by Luigi di Maio, the Five Star Movement received a significant portion of the vote. However, because no party obtained a majority over 40 percent, the Five Star Movement formed a coalition with the right-wing League (*Lega*), which also had a significant number of votes. The League had initially been a northern separatist party but—and through the leadership of Matteo Salvini—moved past corruption scandals, and its association with the Berlusconi era refashioned the League as a right-wing narrative: "Salvini's anti-immigration and anti-EU rhetoric morphed into a wider narrative of defence of the people, of national sovereignty and cultural and religious heritage" (Alcaro and Tocci 2018, 3). Moreover, and crucially, Salvini was masterful in his communication and media strategy: "almost uninterrupted exposure to the public via TV, radio and social media" (Alcaro and Tocci 2018, 4).

5. It is worthwhile to note just how extraordinary and singular was the League's shifting alliance. In the 1990s and 2000s, the League was a regional separatist party that promoted the secession of northern Italy, with lines often drawn right at Florence. Yet it transformed itself into an anti-immigrant and national sovereignty party such that the

foreign national rather than the Italian southerner became the object of unified derision (see D'Alimonte 2019).

6. PD leader Matteo Renzi, representative of the Democractic Party (Partito Democratico, PD) and prime minster of Italy from February 2014 to December 2016, served during a period of transition toward populist rule insofar as he "played on popular resentment against the EU's perceived prevarications (on fiscal policy) and lack of solidarity (on migration)" (Alcaro and Tocci 2018, 2). Yet after his own policies deviated from his political rhetoric and his proposal to amend the Italian constitution failed, Renzi was seen as the very "establishment" he had supposedly aimed to uproot (Alcaro and Tocci 2018, 3).

7. And algorithms are becoming beautiful. A 2014 exhibit called "The Art of the Algorithm" was part of the 2014 London Design Festival and ran with the tag line "Almost every part of our lives, from medicine to music, is now shaped, informed or controlled in some way by algorithms. They have become one of the most powerful forces shaping the twenty-first century, but remain invisible and impenetrable to all but a few." The exhibit featured the work of various artists including Italian artist Carlo Zapponi; he actually calls himself a "maker" and "data visualization designer." Zapponi designs elaborate and beautiful visualizations of data, including funding for Italian political parties, global migration, and sorting algorithms.

8. Grillo was convicted of manslaughter after a 1981 car accident in which three passengers died, and so he cannot be elected to a position. The Five Star Movement bans anyone with a conviction from serving in the party (Hooper 2013). Grillo has been sued several times for defamation.

9. Pseudonym by author; the entire interview of "Sofia Cingolani" was conducted in Italian in March 2019 and translated by author.

10. In *War, Politics, and Superheroes*, Marc DiPaolo (2011) examines how American superheroes and villains reflect contemporary political ideology in the United States. Romero's *Night of the Living Dead*, he suggests, emerged out of a social context of "racism, class warfare, and American imperial policies" (DiPaolo 2011, 250). He examines how Barack Obama's presidential term has led to proliferating narratives of zombie apocalypse. DiPaolo argues that zombie narratives give Americans a catharsis as "the physical manifestation of cultural anxieties about a number of abstract issues." In short, "they give angry Americans something to shoot at" (DiPaolo 2011, 252).

11. He added: "Constantly saying that immigrants bring disease is baseless and puts technical instances in a difficult position, as they are forced into a kind of self-censure in order not to contradict the political field." President of ISS since 2014, Ricciardi had faced controversy after making certain vaccines obligatory in 2017, even though it aligned Italy with international vaccine standards. He was exposed in Giulia Innocenzi's book *Vacci-Nazione* for serving as a consultant for pharmaceutical lobbies for companies such as AbbVie, Novartis, and Pfizer (Innocenzi 2017). In his closing statement, Ricciardi referenced Trump's position of the term "evidence-based science" and added, "It is a particular attitude taken up by populists, who have trouble dealing with science [*hanno grande difficoltà a interagire con la scienza*]" ("Walter Ricciardi" 2019).

12. Casaleggio ran and lost for local office and had political aspirations. He approached Grillo after a show in 2004; Grillo is said to have described him as an "evil genius" and

said he was a modern-day Saint Francis of Assisi but "instead of talking to the birds, he spoke to the internet" (qtd. in Loucaides 2019, 87). After his death in 2019, his son, Davide, took over Casaleggio Associates.

13. Casaleggio was also known for his quirks, like his use of neurolinguistic programming which was purported to mirror client wishes and sway thinking. He also organized psychology sessions for his workers, and spurred Y2K panic by preparing employees for the apocalypse (Loucaides 2019, 86). He also wrote articles that praised Genghis Khan's random murders as a model for "compet[ing] on the web" (Loucaides 2019, 86). He was also criticized for a 2008 video in which he discusses a global conspiracy in which Google governs the world and operates under the power of the Masons and the Bilderberg Group (Loucaides 2019, 90). It is worth saying here that this same spirit of conspiracy and behind-the-curtain way of thinking, in which apparent truths were assumed to be fictions, was cultivated and diffused through Berlusconi's era of mediatized politics. Cynicism and conspiratorial thinking surged in Italy in part because of the entrenchment of a political system that controlled the media and thus was always partial spectacle, aligned with right-wing interests. Casaleggio certainly exerted a right-wing influence on Grillo and on the Five Star Movement, as he was a Euro-skeptic—Five Star allied with Nigel Farage's United Kingdom Independence Party in 2014—and shared anti-immigration views, long before Five Star partnered with the right-wing League in 2018. Before he died in 2016 and left Casaleggio Associates to his son, Davide, he was said to have legally maintained all control over Rousseau, thus taking more control over the Five Star Movement from Grillo, as most of its activity would be on Rousseau rather than Grillo's blog (Loucaides 2019, 93). In fact, Grillo and Casaleggio's last conversation was a phone call fight about this very shift.

14. Rousseau has different nodes or functions, including "Lex iscritti" (Law Suscribers) and "Lex parlamento" (Parliamentary Law).

15. In the early 2010s, M5S had proposed a referendum to end Italy's Eurozone membership. After 2013, M5S slowly tempered this proposal to remain Euro-skeptic but without a more radical proposal to end use of the Euro or EU membership (Alcaro and Tozzi 2018, 3).

16. Di Maio's own recollection of his first election as party representative included a brief speech in which he simply said, "I will no longer call the deputies honorable" (Minuz 2018).

17. Using algorithms and predictive analytics will extend to the management of disasters and crises: "Disasters themselves become a form of 'datafication,' revealing the existence of poor modes of self-governance" (Chandler 2015, 844). The logic of algorithmic prediction, therefore, means that various problems—from slower economies to terrorist attacks to power outages—will be understood as failures to properly calibrate the Big Data system to predict where problems might arise. At a fundamental level, algorithms are machines of prediction, of forecasting, of creating a meaningful relationship between one thing and another, coding input to output. In science and computing I could offer countless examples. In earthquake science, for instance, the M8 is an earthquake prediction algorithm (Molchan and Romashkova 2013).

4. The Trial against Disinformation

1. The victims were mostly students and young people (18 to 29 years old), who accounted for 23 percent of fatalities, and elderly people over age 65, who were 19 percent of fatalities (Calandra 2018).

2. In an attempt to quell the international outrage, Judge Marco Billi declared, "This is not about putting science on trial" ("Sisma L'L'Aquila, sentenza Grandi Rischi: 'Affermazioni approssimative e inefficacy" 2013), *Il Gazettino*, January 18, 2013). In November 2013, the scientists who were found guilty were still in the process of appealing and hoping to reduce prison terms from six years. And while scientists, not science, were technically on trial, this phrase fit with the public discourse that attempted to reframe the trial as stemming from antiscience or credulous beliefs.

3. It was Alan Leshner. Some, however, saw through the hype and recognized the errors of the scientists. For example, American earth scientist Thomas Jordan said that the scientists had flawed logic because shocks are more correlated with the presence of swarms than the absence of swarms.

4. During the trial, in October 2011, witnesses from the town came to give their testimony of the events, including Linda Giugno, who told the jury that the last words she heard from her brother Luigi were "Relax. I'll see you tomorrow" (Cartlidge 2012, 184). He said he didn't think he was in any danger because he had heard the experts on television tell him so. He did not wake his wife, who was pregnant and scheduled to deliver that day, or their two-year-old son.

5. The judges reasoned that the earthquake was not "the exclusive cause of the death or the injury . . . but the imprudent behavior of the Commission members" (Cirillo 2013).

6. Moreover, the mayor, Massimo Cialente, was quoted on a twenty-four-hour news network as saying, "There should be absolutely no risk" of building collapse (Cartlidge 2012, 185).

7. Žižek (2006) also suggests: "While the view of scientific discourse as involving a pure description of facticity is illusory, the paradox resides in the coincidence of bare facticity and radical voluntarism: facticity can be sustained as meaningless, as something that 'just is as it is,' only if it is secretly sustained by an arbitrary divine will" (164). For Žižek, Christian theology and the belief in God were actually essential to the development of modern science in this "voluntarism," which suggests that rational facts rely on the will of God. Ciccozzi (2016) also describes the trial in L'Aquila as a ritual that granted "an aura of sacredness to science" (72).

8. The Commission made "approximate, generic, and ineffective claims" and reassured Aquilans regarding "the predictability of earthquakes, seismic precursors, the development of the coming earthquake swarm, the normality of the phenomenon, the release of energy caused by the earthquake swarm as a favorable situation" ("Sisma" 2013). The commission, reads the verdict, "could have intervened in the victims' volitional processes (*processi volitivi*)" ("Sisma" 2013).

9. In *Semblance and Event*, Brian Massumi (2011) theorizes events: "Action is only half the event: action-reenaction; rhythm-reverberation; point, virtual counterpoint. . . . In every event of perception, there is a differential co-involvement in the dynamic unity of one and the same occurrence. . . . Every movement has an activation contour, a rhythm

of activity: vitality affect" (115). He also shares, "The suspension-event is an incorporeal envelope of sociality" (35).

10. Only thirty-five events in Italy have been declared "big events" since the inception of the law, half of which were religious or Vatican-related events, like a visit by Pope Benedict to Genoa.

11. Calandra (2018) makes an interesting point that the new housing units were built in linear fashion, along roads, which was a new urban geography in contrast to the concentric circles that represented the hallmark of a medieval town center. The new and foreign layout of housing contributed to the alienation of residents.

12. The bad-luck throwers were seen not as diabolical but rather unwillingly or unknowingly had this power of expulsion (de Ceglia 2011, 80). This is why I adopt the notion of throwing or tossing *jetta*: because there is a differentiation between the man and the substance or "emanations" or "emissions" (Dennert qtd. in de Ceglia 2011, 81). I also use the spelling that de Ceglia used; other variations include *jattatura*.

13. Cinematic star Totò played the protagonist in the 1954 film based on Pirandello's work, *This Is Life* (*Questa é la vita*). The play ends with Chiarchiaro's collection of fees without his official license.

14. De Martino has a unique sensibility toward the experience of ritual and belief, but ultimately, he viewed Neapolitans as a "failure of rationalism" (Magliocco 2012, 14) and imagined Southern Italy as necessarily partially embedded in pre- and post-Enlightenment discourses of rationality and reason. His phrase "crisis of presence" also implied a kind of lack of agency and a subject who had the constant desire to protect oneself and forestall danger.

15. The Comaroffs have called the judiciaries sites of "enchanted displacement" because of the persistent belief that courts can produce "social harmony" (Comaroff and Comaroff 2000, 328). The L'Aquila court, then, was called on to settle a question of reason and unreason and punish what would exceed a rational actor's assessment of risk, in ways that seem to transport the fantasy of order to the court.

16. Pirandello seems to have been playing with this in his play as well. The judiciary mediated between the logics of prediction and risk estimation, yet, all the while, Italy's apparent entrenchment in superstition and supernatural belief seems to have haunted this trial.

17. Sandro Bondi, the minister of culture, boycotted the Cannes and called the film "propaganda offending the truth and the whole Italian people" (Pisa 2010).

18. For example: private hotels, private construction and design firms, essentially doling out funds to contractor friends. Various forms of work and reconstruction—including allowing citizens to safely reenter their homes and collect their belongings—were stopped for at least a month (Cerasoli 2010, 41).

19. In *Semblance and Event*, Brian Massumi (2011) theorizes events: "Action is only half the event: action-reenaction; rhythm-reverberation; point, virtual counterpoint. . . . In every event of perception, there is a differential co-involvement in the dynamic unity of one and the same occurrence. . . . Every movement has an activation contour, a rhythm of activity: vitality affect" (115). He also shares, "The suspension-event is an incorporeal envelope of sociality" (35).

20. See note 10 above.

21. Another call recorded two men with ties to Civil Protective services speaking only hours after the earthquake and saying, "We must act immediately."

5. Scientific Anesthetization in the Anthropocene

1. The study of neutrino oscillations, and the OPERA project more broadly, has significance in contributing to understandings of dark matter and glaciers to hadron therapy ("OPERA News and Updates" 2018).

2. Bauer and Bhan (2016) suggest that the human and nature binaries persist in scholarly analysis and discussion on the Anthropocene. They suggest that political welfare still rests on a division between nature and society and also human and nonhuman: "[the Anthropocene] fails to question the enduring legacies and implications of separating human histories from natural histories" (63).

3. While the United States is famously the site of skepticism and deliberate disinformation regarding climate change, this kind of climate change denial is far rarer in Italy. However, the kind of doubt that remains even with a public that accepts the scientific consensus on climate change is still dubious toward the predicted effects of climate change, as they are empirically challenging to predict.

4. Giuliani has since been invited to the American Geophysical Union to present his work to its members in San Francisco and will serve as part of a panel working toward a seismic early warning system (Dollar 2010).

5. "Dietrologia" has also been translated as "behindology" (G. 2011; Hooper 2016, 69). Hooper (2016) uses behind-ism (69–70).

6. They create this using HAARP (High Frequency Auroral Research Program), a secret, very low frequency radio wave generation which the narrator calls "like a billion watt microwave" that alters the ionosphere. They "adjust the vibration underneath the rock" or "heat up subterranean water."

Conclusion: Mirrored Window World

1. In fact, the mirror also has symbolic resonance for the history of the Five Star Movement. Casaleggio, the famed creator of Rousseau, Five Star's digital platform, had been a figure in computer science in Italy for years before partnering with Grillo. His former company Webegg experimented with neurolinguistic programming in the late 1990s, which would "influence people by surreptitiously tapping into the unconscious patterns that guide their behavior" and "'mirror' that client" (Loucaides 2019, 86). Thus, the programmers were already adopting methods to invisibly shift thinking and model behaviors on a client's already existing taste and preferences—it was a precursor to what we now call algorithmic technologies like filter bubbles and personal tracking.

2. My move to historicize prediction and causality has anthropological precedents. Jane Guyer (2007) suggests that "the near future is a kind of hiatus" that converges in

evangelical time and in neoliberal economic time: "The individual and collective near future [is] thinned out of its complexity as a theory or doctrine embodied in guidelines and benchmarks and indexed to a defined, more distant collective future" (416). Guyer finds odd partners in this kind of temporal framework between monetary theorists and fundamentalist Christians. She suggests that the ideologies converge on a kind of "date-as-event" organization (417), which also works to diminish attention to the immediate past and proximate future. Thus, she argues, temporal engagement means a focus on "punctuated time," which is "event-driven," whether that means a second coming or a debt schedule (416). Just as Guyer examines how a convergence of ideologies creates punctuated time, I am suggesting that the convergence of political disinformation, human-created natural disaster, and algorithmic culture makes prediction appear simultaneously mystified and objective.

3. Eli Pariser (2011) has also called the Google algorithm "a kind of one-way mirror, reflecting our own interests while algorithmic observers watch you click" (3).

4. One Italian tech writer describes how the algorithms of music programs like Spotify have become carefully calibrated to what users "like" to generate income (Dawson 2015). Still, the underlying model has worked to generate both quality playlists and user satisfaction such that an algorithm seems preferable to a friend's suggestions.

5. Sunstein (2017) imagines how a 1990s prophecy about a "Daily Me," a personally tailored newspaper, has in some ways been produced via personalized algorithms (2–4).

6. Bogost illustrates this well by likening algorithms to the metaphor of manufacturing: just as the so-called automated manufactured rubber ducky masks "vinyl plastic, injection molding, the hands and labor of Chinese workers, the diesel fuel of ships and trains and trucks, the steel of shipping containers," so, too, does the Google algorithm hide "a confluence of physical, virtual, computational, and non-computational stuffs—electricity, data centers, servers, air conditioners, security guards, financial markets."

7. Striphas (2015) argues algorithms "changed how the category culture has long been practiced, experienced and understood" (396).

8. In a similar vein, Leon Weseltier (2015) suggests part of the magical appearance of objectivity rests on numerical processes: "[Quantification] is enabled by the idolatry of data, which has itself been enabled by the almost unimaginable data-generating capabilities of the new technology."

9. While Seaver argues algorithms should be viewed as not "autonomous technical objects but complex sociotechnical systems," his work also shows us how the human labor and human design can easily slip beneath the radar and become invisible in how users and citizens understand the technology they use (378). Seaver (2018) warns that anthropologists not consider algorithms as a part of a quasi "savage slot," or, as he dubs it, the "analog slot . . . full of the wild and distinctively human remainders that computing can supposedly never consume: our sensitivity to context, our passions and affects, our openness to serendipity and chance, our uncanny navigation of the cultural field" (380). It is a sage warning against a dangerous and unproductive idealization of analog times, a nostalgic and simplified "before" computing and algorithmic dominance in everyday life. Because my principal concern is the misunderstandings and mystification of algorithms by everyday citizens, Seaver's concern actually supports the same idea. Yet, for him, rei-

fied understandings of algorithms extend not just to everyday citizens but also to cultural analysts.

10. Moreover, the process seems magical because of an apparently seamless cause-effect: "The magic plays out whenever you order an item online only to see its physical embodiment at your doorstep a few days later" (Ekbia and Nardi 2018, 362).

11. Besteman (2019) argues that roboprocesses are "non-transparent and difficult to comprehend; those who write and deploy them are not accountable for their effects; they are secret." They represent a different form of modernity, that is, other forms of information processing and automation (6).

12. Hugh Gusterson (2019) argues, "The conjuncture between computerization and neoliberalism has produced roboprocesses skewed in favor of corporate profit-making, mass surveillance, and the re-entrenchment of racial and class-based inequalities" (13).

13. David McNally (2011) proposes critical attention to these occult imaginings as a way to develop the "dialectical optics, ways of seeing the unseen" in order to examine this kind of capitalism, precisely because so much of global capitalism—the slave labor and bondage, the hidden movement of people, commodities, and material things, the abject poverty and displacement—are often unseen (6).

14. There is also a somatic aspect to these economic disappearance acts as David McNally, after Scheper-Hughes (1996) and Comaroff and Comaroff (2000), attends to "tales of bodysnatching of abduction, ritual murder, and organ theft [that] traverse folklore, science fiction, film, video, and print media" (McNally 2011, 3). The hypercommodification of the body gives rise to perverse fears of bodily alienation, as McNally (2011) puts it, "monstrous dislocations at the heart of commodified existence" (8). Thus, algorithms, in this sense and in their role in driving production and consumption, represent a kind of interconnecting water system of capitalism's "cartography of the invisible" (McNally 2011, 7).

15. *Le Iene* (The hyenas) is a popular political talk show in Italy. It refers indirectly to the title of Quentin Tarantino's film *Reservoir Dogs,* which was titled *Le Iene* in Italian.

Bibliography

Abu El-Haj, Nadia. 2012. *The Genealogical Science: The Search for Jewish Origins and the Politics of Epistemology*. Chicago: University of Chicago Press.

Adas, Michael. 1990. *Machines as the Measure of Men: Science, Technology, and Ideologies of Western Dominance*. Ithaca, NY: Cornell University Press.

Agamben, Giorgio. 2002. "Security and Terror." Translated by Caroline Emcke. *Theory and Event* 5 (4): 1–6. https://philpapers.org/rec/AGASAT-3.

——. 2005. *State of Exception*. Translated by Kevin Attell. Chicago: University of Chicago Press.

——. 2015. "Security City." In *Michele Nastasi, Suspended City: L'Aquila after the Earthquake*, edited by Maddalena D'Alfonso, 9. New York: Actar.

AGI. 2012. Mediaset: "Striscia la notizia' il programma piu" visto. Prima Online, August 6. http://www.primaonline.it/2012/06/08/107068/mediaset-striscia-la-notizia-il-programma-piu-visto-2/.

Agnew, John, and Michael Shin. 2016. "Electoral Dramaturgy: Insights from Italian Politics about Donald Trump's 2015–16 Campaign Strategy . . . and Beyond." *Southeastern Geographer: Special Forum on the Geographies of the 2016 Presidential Election* 53 (3): 265–272.

——. 2017. "Spatializing Populism: Taking Politics to the People in Italy." *Annals of American Association of Geographers* 107 (4): 915–933. https://doi.org/10.1080/24694452.2016.1270194.

Albini, Andrea. 2007. "La Radioestesia fallisce un test in doppio cieco." *Scienza & Paranormale* 74 (August). http://www.cicap.org/new/articolo.php?id=273318.

Alcaro, Riccard, and Nathalie Tocci. 2018. "Italy's Unruly Rulers." *Istituto Affari Internazionali Commentaries* 18 (70): 1–5.

Alderman, Liz. 2013. "On the Brink in Italy." *New York Times*, March 11, 2013. https://www.nytimes.com/2013/03/12/business/global/12iht-euitaly12.html.

Alemanno, Alberto, and Kristian Cedervall Lauta. 2014. "The L'Aquila Seven: Re-Establishing Justice after Natural Disaster." *European Journal of Risk Regulation* 2:137–145. https://papers.ssrn.com/sol3/papers.cfm?abstract_id=2461327.

Allum, Felia. 2011. "Silvio Berlusconi and His 'Toxic' Touch." *Representation* 47 (3): 281–294. https://doi.org/10.1080/00344893.2011.596434.

Amodeo, Francesco. 2019. *La Matrix Europea: Il piano di conquista del Cartello Finanziario in Italia.* Francesco Amodeo Matrix Edizioni.

ANSA. 2013. "Letta, linguaggio sovversiva verita." April 29, 2013. http://www.ansa.it/web/notizie/rubriche/topnews/2013/04/29/Letta-linguaggio-sovversivo-verita-_8629401.html.

Arcovio, Valentina. 2012. "Tutte le bufale scientifiche di Beppe Grillo: Il leader del Movimento 5 Stelle ne ha dette di cotte e di crude, specie in campo scientifico." *Wired*, May 18, 2012. http://daily.wired.it/news/scienza/2012/05/18/bufale-scientifiche-beppe-grillo-23666.html.

Arcuri, Alessandra, and Marta Simoncini. 2015. "Scientists and Legal Accountability: Lessons from the L'Aquila Case." *EUI Working Paper Law*. Badia Fiesolana: European University Institute, 1–17.

Ardizzoni, Michela. 2005. "Redrawing the Boundaries of Italianness: Televised Identities in the Age of Globalisation." *Social Identities* 11 (5): 509–530. https://doi.org/10.1080/13504630500408123.

——. 2009. "Democracy without Dissent: Satirical News in Italy." *FlowTV*, July 11. http://flowtv.org/2009/07/democracy-without-dissent-satirical-news-in-italymichela-ardizzoni-university-of-colorado-boulder/.

Arnold, David. 2013. *Everyday Technology: Machines and the Making of India's Modernity*. Chicago: University of Chicago Press.

Bader, Andrea, Joseph O. Baker, and Andrea Molle. 2012. "Countervailing Forces: Religiosity and Paranormal Belief in Italy." *Journal for the Scientific Study of Religion* 51 (4): 705–720. https://doi.org/10.1111/j.1468-5906.2012.01674.x.

Bailey, Michael. 2007. *Magic and Superstition in Europe: A Concise History from Antiquity to the Present*. Plymouth, UK: Rowman & Littlefield.

Bakhtin, Mikhail. 1984. *Rabelais and His World*. Translated by Hélène Iswolsky. Bloomington: Indiana University Press.

Balmer, Crispian. 2017. "Silvio Berlusconi Says His Favourite Thing about Donald Trump Is His Wife Melania." *Independent*, June 23, 2017. https://www.independent.co.uk/news/world/americas/us-politics/trump-berlusconi-melania-favourite-thing-italy-us-president-comments-a7805676.html.

Baraniuk, Chris. 2017. "The Animals Thriving in the Anthropocene." BBC. http://www.bbc.com/future/story/20170801-the-animals-thriving-in-the-anthropocene.

Barberi, Franco et al. 2009. "Major risks committee minutes L'Aquila".

L'espresso, March 31, 2009. http://www.seis.nagoya-u.ac.jp/yamaoka/iweb/NU-site/LAquila_files/cgr-english.pdf

Barcellona, Gaia Scorza. 2016. "L'eredità di Casaleggio: Come funziona Rousseau, 'sistema operativo' del M5s." *La Repubblica*, April 13, 2016. https://www.repubblica.it/tecnologia/2016/04/13/news/l_eredita_di_gianroberto_casaleggio_arriva_rousseau_sistema_operativo_per_m5s-137514863/#gallery-slider=137522242.

Barrios, Roberto E. 2017. "What Does Catastrophe Reveal for Whom? The Anthropology of Crises and Disaster at the Onset of the Anthropocene." *Annual Review of Anthropology* 46:151–166. doi:10/1146/annurev-anthro-102116-041635.

Bauer, Andrew M., and Mona Bhan. 2016. "Welfare and the Politics and Historicity of the Anthropocene." *South Atlantic Quarterly* 115 (1): 61–87. https://doi.org/10.1215/00382876-3424753.

Baym, Geoffrey, and Jeffrey P. Jones. 2012. "News Parody in Global Perspective: Politics, Power, and Resistance." *Popular Communication* 10:2–13. doi:10.1080/15405702.2012.638566.

Beck, Estee N. 2015. "The Invisible Digital Identity: Assemblages in Digital Networks." *Computers and Composition* 35 (March 2015): 125–140. https://doi.org/10.1016/j.compcom.2015.01.005.

Bei, Francesco. 2011. "Il Cavaliere nel fortino di Palazzo Chigi." *La Repubblica*, June 17. http://www.repubblica.it/politica/2011/06/17/news/berlusconi_mollare-17806370/.

Benadusi, Mara. 2016. "The Earth Will Tremble? Expert Knowledge Confronted after the 2009 L'Aquila Earthquake." In *Archivio Antropologico Mediterraneo Online*. Mara Benadusi and Sandrine Revet, eds. 18 (2): 17–32.

Benadusi, Mara, and Sandrine Revet. 2016. "Disaster Trials: A Step Forward." In *Archivio Antropologico Mediterraneo online*, edited by Mara Benadusi and Sandrine Revet, 18 (2): 7–16.

Bendoris, Matt. 2007. "Who Really Did for Diddy Dida?" *Sun* (London), October 6.

Benini, Stefania. 2012. "Televised Bodies: Berlusconi and the Body of Italian Women." *Journal of Italian Cinema and Media Studies* 1 (1): 87–102. https://doi.org/10.1386/jicms.1.1.87_1.

Berardi, Franco, and Marco Jacquemet. 2009. *Ethereal Shadows: Communications and Power in Contemporary Italy*. Chico, CA: AK Press.

Berberi, Leonard. 2015. "Antropocene: 1610, l'anno in cui l'uomo cambiò il pianeta." *Corriere della Sera*, March 15. http://www.corriere.it/scienze/15_marzo_13/antropocene-era-geologica-umana-24050fdc-c968-11e4-84dd-480351105d62.shtml.

"Berlusconi, 'Occidente, civilta' superior." 2001. *Corriere della Sera*, September 27, 2001. https://www.corriere.it/Primo_Piano/Esteri/09_Settembre/26/berlusconi.shtml.

Berlusconi, Silvio. 2001. *Una storia italiana*. Milano: Mondadori.

"Berlusconi, 'Occidente, civilta' superior." 2001. *Corriere della Sera*, September 27, 2001. https://www.corriere.it/Primo_Piano/Esteri/09_Settembre/26/berlusconi.shtml.

Besteman, Catherine. 2019. "Afterword: Remaking the World." In *Life by Algorithms: How RoboProcesses Are Remaking our World*, edited by Catherine Besteman and Hugh Gusterson, 1–27. Chicago: University of Chicago Press.

Biagioli, Mario. 1999. "Introduction." In *The Science Studies Reader*, edited by Mario Biagioli, xi–xvi. New York: Routledge.

Bierma, Nathan. 2006. "Talking about the Word of the Year." *Chicago Tribune*, December 27, 2006. https://www.chicagotribune.com/news/ct-xpm-2006-12-27-0612270247-story.html.

Bigi, Alessandro, Kirk Plangger, Michelle Bonera, and Colin L. Campbell. 2011. "When Satire Is Serious: How Political Cartoons Impact a Country's Brand." *Journal of Public Affairs* 11 (3): 148–155. https://doi.org/10.1002/pa.403.

Bijker, Wiebe, and Trevor Pinch. 1987. "The Social Construction of Facts and Artifacts. In *The Social Construction of Technological Systems*, edited by B. Wiebe, T. Hughes, and T. Pinch, 17–50. Boston, MA: MIT Press.

Billig, Michael. 2001. "Humour and Embarrassment: Limits of 'Nice-Guy' Theories of Social Life." *Theory, Culture and Society* 18 (23): 23–43. https://doi.org/10.1177/02632760122051959.

——. 2005. *Laughter and Ridicule: Towards a Social Critique of Humour*. London: Sage.

——. 2011. "The Birth of Comedy Italian Style." In *Popular Italian Cinema: Culture and Politics in a Postwar Society*, edited by Flavia Brizio-Skov, 107–152. London: I. B. Tauris.

Bini, Andrea, ed. 2011. "The Birth of Comedy Italian Style." In *Popular Italian Cinema: Culture and Politics in a Postwar Society,* edited by Flavia Brizio-Skov, 107–152. London: I. B. Tauris.

Biorcio, Roberto, and Paolo Natale. 2013. *Politica a 5 stelle: Idee, storia e strategia del movimento di Grillo*. Trebaseleghe, PD: Feltrinelli: Serie Bianca.

Black, Rachel E. 2012. *Porta Palazzo: The Anthropology of an Italian Market*. Philadelphia: University of Pennsylvania Press.

Bobba, Giuliano, and Duncan McDonnell. 2016. "Different Types of Right-Wing Populist Discourse in Government and Opposition: The Case of Italy." *South European Society and Politics* 21 (3): 281–299. https://doi.org/ 10.1080/13608746.2016.1211239.

Bogost, Ian. 2015. "The Cathedral of Computation." *Atlantic*, January 15, 2015. https://www.theatlantic.com/technology/archive/2015/01/the-cathedral-of-computation/384300/.

Bonneuil, Christophe, and Jean-Baptiste Fressoz. 2016. *The Shock of the Anthropocene: The Earth, History and Us*. Translated by David Fernbach. London: Verso.

Bordignon, Fabio, and Luigi Ceccarini. 2013. "Five Stars and a Cricket: Beppe Grillo Shakes Italian Politics." *South European Society and Politics* 10:1–23. https://doi.org/10.1080/13608746.2013.775720.

Boyer, Dominic. 2013. "Simply the Best: Parody and Political Sincerity in Iceland." *American Ethnologist* 40 (2): 276–287. https://doi.org/10.1111/amet.12020.

——. 2013a. *The Life Informatic: Newsmaking in the Digital Era*. Ithaca, NY and London, UK: Cornell University Press.

——. 2018. "Our Post-Post-Truth Condition." *Berliner Blätter* 80 (1): 83–90.

Boyer, Dominic, and Alexei Yurchak. 2010. "American Stiob: Or, What Late-Socialist Aesthetics of Parody Reveal about Contemporary Political Culture in the West." *Cultural Anthropology* 25 (2): 179–221. https://doi.org/10.1111/j.1548-1360.2010.01056.x.

——. 2011. "Stiob Invades Washington (and Reykjavik): How the Aesthetics of Overidentification Are Entering Western Political Culture." Paper presented at the Society for the Anthropology of Europe Mini-Symposium on Politics and Performance, 18th International Conference of Europeanists, Barcelona, June 20–22.

Bozdag, Engin. 2013. "Bias in Algorithmic Filtering and Personalization." *Ethics of Information Technology* 15:209–227. https://doi:10.1007/s10676-013-9321-6.

——. 2015. *Bursting the Filter Bubble: Democracy, Design, and Ethics*. Delft: CPI Koninklijke Wohrmann.

Brewer, Richard. 1992. "A Deficiency of Credulousness." *BioScience* 42 (2): 123–124. https://doi.org/10.2307/1311653.

Briziarelli, Marco. 2011. "Neoliberalism as a State-Centric Class Project: The Italian Case." *Continuum: Journal of Media and Cultural Studies* 25 (1): 5–17. https://doi:10.1080/10304312.2011.538368.

Buonadonna, Marta. 2015. "Salute e ambiente: L'umanità sta meglio, a spese del pianeta." *Panorama*, July 21.

Buonfiglioli, Francesca. 2016. "M5s, le (assured) proposte di leggi degli attivisti su Rousseau." *Lettera 43*, December 28, 2016. https://www.lettera43.it/it/articoli/politica/2016/12/28/m5s-le-assurde-proposte-di-legge-degli-attivisti-su-rousseau/207481/.

Butler, Ella. 2010. "God Is in the Data: Epistemologies of Knowledge at the Creation Museum." *Ethnos: Journal of Anthropology* 75 (3): 229–251. https://doi.org/10.1080/00141844.2010.507907.

Caballero, Adelaida. 2012. "The Haunted Self: Intersubjectivity and Collective Memory in First-Hand Eyewitness Accounts of Paranormal Experiences." Master's thesis, Institutionen for Kulturantropologi och Etnologi, Uppsala Universitet.

Cabot, Heath. 2014. *On the Doorstep of Europe: Asylum and Citizenship in Greece*. Philadelphia: University of Pennsylvania Press.

Calandra, Lina Maria. 2018. "Considerations on the Role of Citizen Participation in L'Aquila (Italy)." In *Governance of Risk, Hazards and Disasters: Trends in Theory and Practice*, edited by Giuseppe Forino, Sara Bonati, and Lina Maria Calandra, 65–78. London: Routledge. https://doi.org/10.4324/9781315463896.

Calcutt, Andrew. 2016. "The Truth about Post-Truth Politics." *Newsweek*, November 21. http://www.newsweek.com/truth-post-truth-politics-donald-trump-liberals-tony-blair-523198.

Callicot, J. Baird. 2015. "Science as Myth (Whether Sacred or Not), Science as Prism." *Journal for the Study of Religion, Nature and Culture* 9 (2): 155–168. https://doi.org/10.1558/jsrnc.v9i2.27264.

Cameron, Euan. 2005. *Interpreting Christian History: The Challenge of the Churches' Past*. Malden, MA: Blackwell Publishing. https://doi:10.1002/9780470774168.

Campus, Donatella. 2010. "Mediatization and Personalization of Politics in Italy and France: The Cases of Berlusconi and Sarkozy." *International Journal of Press/Politics* 15 (2): 219–235. https://doi.org/10.1177/1940161209358762.

Caporale, Giuseppe. 2014. "Terremoto dell'Aquila, Enzo Boschi." *La Repubblica*, November 11, http://www.repubblica.it/cronaca/2014/11/11/news/terremoto_dell_aquila_enzo_boschi_noi_scienziati_siamo_stati_usati_io_la_gente_non_l_avrei_rassicurata-100281289/?refresh_ce.

Cartlidge, Edwin. 2012. "Aftershocks in the Courtroom." *Science* 336 (October 12): 184–187. https://doi.org/10.1126/science.338.6104.184.

Caruso, Loris. 2016. "Economia e Populismo: Il traversalismo del Movimento 5 Stelle alla prova della dimensione economia sociale." *Quaderni di Scienza Politica* 23: 25–54.

——. 2017. "Digital Capitalism and the End of Politics: The Case of the Italian Five Star Movement." *Politics and Society* 45 (4): 585–609. https://doi.org/10.1177/0032329217735841.

Casaleggio, Gianroberto. 2001. *Il Web è morto, viva il Web.* Web Marketing Tools libri.

Casalini, Simona. 2016. "M5S e l'algoritmo contro i traditori." *La Repubblica*, May 19, http://www.repubblica.it/speciali/politica/elezioni-comunali-edizione2016/2016/05/19/news/algoritmo_comico_battuta_di_maio_raggi-140137542/.

Castells, Manuel. 1996. *The Rise of the Network Society.* Malden, MA: Wiley.

——. 2010. *The Information Age: Economy, Society, and Culture.* West Sussex, UK: Wiley Blackwell.

Cazzullo, Aldo. 2011. "Inutile attaccare Silvio." *Corriere della Sera*, September 21, 2011. https://www.corriere.it/politica/11_settembre_21/ricci-berlusconi-intrvista-addio-cazzullo_bb48fa0c-e410-11e0-bb93-5ac6432a1883.shtml.

Cepernich, Christopher. 2008. "Landscapes of Immorality: Scandals in the Italian Press (1998–2006)." *Perspectives on European Politics and Society* 9 (1): 95–109. https://doi.org/10.1080/15705850701825568.

Cerasoli, Domenico. 2010. "De L'Aquila non resta che il nome: Racconta di un terremoto." *L'Aquila: Dietro la catastrophe; Meridiana* 55–56:35–58. https://www-jstor-org.proxy.library.nyu.edu/stable/23204198.

Chakrabarty, Dipesh. 2012. "Postcolonial Studies and the Challenge of Climate Change." *New Literary History* 43 (1): 1–18. https://muse.jhu.edu/article/477475.

Chandler, David. 2015. "A World without Causation: Big Data and the Coming of Age of Posthumanism." *Millennium: Journal of International Studies* 43 (3): 833–851. https://doi.org/10.1177/0305829815576817.

Cheney-Lippold, John. 2011. "A New Algorithmic Identity: Soft Biopolitics and the Modulation of Control." *Theory, Culture & Society* 28 (6): 164–181. https://doi:10.1177/0263276411424420.

Chow, Denise. 2017. "In Photos: The Best Signs from the 2017 March on Science." *LiveScience*, April 22, 2017. https://www.livescience.com/58791-2017-march-for-science-photos.html.

Chu, Henry. 2012. "Will Italy's Sexism Get the Boot?" *Chicago Tribune*, April 8, 2012.

CICAP. 2011. "Venerdi 17: Tutti alla giornata anti superstizione con il CICAP," June 16, 2011. http://www.cicap.org/new/articolo.php?id=274516.

Ciccozzi, Antonello. 2009. "Aiuti e miracoli ai margini del terremoto de L'Aquila." *Meridiana* 65–66, *L'Aquila 2010: Dietro La* Catastrofe, 227–255. http://www.jstor.org/stable/23204207.

——. 2016 "Forms of Truth in the Trial against the Commission for Major Risks—Anthropological Notes." In *Archivio Antropologico Mediterraneo online*, edited by Mara Benadusi and Sandrine Revet. 18 (2): 65–82.

Cirillo, Giovanni. 2013. "Terremoto L'L'Aquila, 'I 7 esperti condannati? Perche rassicurarano i cittadini." *Il Fatto Quotidiano*, September 24. http://www.ilfattoquotidiano.it/2013/09/24/terremoto-lL'Aquila-7-della-commissione-condannati-perche-rassicurarono-cittadini/721220/.

City. 2010. "Superstizione, italiani quasi sul tetto d'Europe." June 21. http://city.corriere.it/2010/06/22/milano/documenti/superstizione-italiani-quasi-tetto-d-europa-20816083464.shtml.

Clementi, Marco, David G. Haglund, and Andrea Locatelli. 2017. "Making America Grate Again: The 'Italianization' of American Politics and the Future of Transatlantic Relations in the Era of Donald J. Trump." *Political Science Quarterly* 132 (3): 495–525. https://doi.org/10.1002/polq.12655.

"Cloud-seeding e chemtrails." 2008. Sciechimiche.org, June 22. http://www.sciechimiche.org/scie_chimiche/index.php?option=com_content&task=view&id=589&Itemid=1.

Cohen, Roger. 2016. "The Trump-Berlusconi Syndrome." *New York Times*, March 14, 2016. https://www.nytimes.com/2016/03/15/opinion/the-trump-berlusconi-syndrome.html.

Collier, Stephen J., and Andrew Lakoff. 2015. "Vital Systems Security: Reflexive Biopolitics and the Government of Emergency." *Theory, Culture & Society* 32 (2): 19–51. https://doi.org/10.1177/0263276413510050.

Comaroff, Jean, and John Comaroff. 2000. "Millennial Capitalism: First Thoughts on a Second Coming." *Public Culture* 12 (2): 291–343. https://muse.jhu.edu/article/26196.

Concina, Michele, and Alberto Costa, eds. 2001. *Una Storia Italiana*. Milan: Mondadori Printing S.p.A.

Conversi, Daniele, and Luis Moreno. 2017. "Antropocene, il nuovo mondo che finisce." *La Repubblica MicroMega*. http://temi.repubblica.it/micromega-online/antropocene-il-nuovo-mondo-che-finisce/?refresh_ce.

Corriere della Sera. 2012. "Striscia, medico indagato per violenza sessuale." *Corriere della Sera*, May 17. http://www.corriere.it/cronache/12_maggio_17/striscia-medico-violenza-indagato_b44299c8-a03c-11e1-bef4-97346b368e73.shtml.

Cosentino, Gabriele. 2012. "The Comical Inquisition: *Striscia la Notizia* and the Politics of Fake News on Italian Television." *Popular Communication* 10:52–65. https://doi.org/10.1080/15405702.2012.638570.

———. 2017. *L'era del post-verità: Media e populismi dalla Brexit a Trump*. Imprimatur Editore.

Couldry, Nick. 2014. "Inaugural: A Necessary Disenchantment: Myth, Agency and Injustice in a Digital World." *Sociological Review* 62:880–897. https://doi.org/10.1111/1467-954X.12158.

———. 2017. "Surveillance-democracy." *Journal of Information Technology & Politics* 14 (2): 182–188. https://doi.org/10.1080/19331681.2017.1309310.

Couldry, Nick, and Ulises A. Mejias. 2019. "Data Colonialism: Rethinking Big Data's Relation to the Contemporary Subject." *Television & New Media* 20 (4): 336–349. https://doi.org/10.1177/1527476418796632.

Critchley, Simon. 2012. *The Faith of the Faithless: Experiments in Political Theology*. London: Verso.

Cronon, William. 1990. "Modes of Prophecy and Production: Placing Nature in History." *Journal of American History* 76 (4): 1122–1131. http://www.jstor.org/stable/2936590.

Csordas, Thomas. 2007. "Global Religion and the Re-enchantment of the World: The Case of the Catholic Charismatic Renewal." *Anthropological Theory* 7 (3): 295–314. https://doi:10.1177/1463499607080192.

Curtis, Neal. 2013. "Thought Bubble: Neoliberalism and the Politics of Knowledge." *New Formations* 80–81:73–81. https://doi.org/10.3898/NEWF.80/81.04.2013.

Cuzzocrea, Annalisa. 2017. "Nuovo attacco hacker a Rousseau." *La Repubblica*, August 4, 2017. https://www.repubblica.it/politica/2017/08/04/news/nuovo_attacco_hacker_a_rousseau_facile_giocare_con_i_vostri_voti_-172309654/.

Dahlgren, Peter, and Claudia Alvares. 2013. "Political Participation in the Age of Mediatisation: Towards a New Research Agenda." *Journal of the European Institute for Communication and Culture* 20 (2): 47–66. http://doi: 10.1080/13183222.2013.11009114.

D'Alimonte, Roberto. 2019. "How the Populists Won in Italy." *Journal of Democracy* 30 (1): 114–127. https://doi.org/10.1353/jod.2019.0009.

Dalla Casa, Stefano. 2012. "Il complesso di Cassandra." *Oggi Scienza*, May 30. http://oggiscienza.it/2012/05/30/il-complesso-di-cassandra/.

Davies, Lizzy. 2013. "Berlusconi Found Guilty after Case that Cast Spotlight on Murky Premiership." *Guardian,* June 25, 2013. https://www.theguardian.com/world/2013/jun/24/silvio-berlusconi-guilty-underage-prostitute.

Davies, William. 2019. *Nervous States: Democracy and the Decline of Reason*. New York: W.W. Norton.

Dawson, Oliver. 2015. "Meglio il consiglio di un amico, o quello di un algoritmo?" *Rockit.it*. https://www.rockit.it/articolo/algoritmi-consigli-spotify.

De Ceglia, Francesco Paolo. 2011. "'It's Not True, but I Believe It': Discussions on jettatura in Naples between the End of the Eighteenth and Beginning of the Nineteenth Centuries." *Journal of the History of Ideas* 72 (1): 75–97. https://muse.jhu.edu/article/413475.

De Gregorio, Concita. 1992. "C'era una volta la corte di 'nani e ballerine.'" *La Repubblica*, November 25, 1992, 8. http://ricerca.repubblica.it/repubblica/archivio/repubblica/1992/11/25/era-una-volta-la-corte-di.html.

De Martino, Ernesto. 2004 [1959]. *Sud e magia*. Milan: Feltrinelli Editore.

Deleuze, Gilles. 1971. *Masochism: An Interpretation of Coldness and Cruelty*. Translated by Jean McNeil. New York: George Braziller.

Dendle, Peter. 2007. "The Zombie as Barometer of Cultural Anxiety." In *Monsters and the Monstrous: Myths and Metaphors of Enduring Evil*, edited by Niall Scott, 45–59. Amsterdam: Editions Rodopi.

Di Martino, Loredana. 2011. "From Pirandello's Humor to Eco's Double Coding: Ethics and Irony in Modernist and Postmodernist Italian Fiction." *MLN: Modern Language Notes* 126 (1): 137–156. https://doi:10.1353/min.2011.0002.

Dillon, Michael. 2007. "Governing Terror: The State of Emergency of Biopolitical Emergence." *International Political Sociology* 1:7–28.

DiPaolo, Marc. 2011. *War, Politics, and Superheroes: Ethics and Propaganda in Comics and Film*. Jefferson, NC: McFarland.

Dmitriev, Anatolii Vasil'evich. 2006. "Humor and Politics." *Anthropology and Archaeology of Eurasia* 44 (3): 64–100. https://doi.org/10.2753/AAE1061-1959440304.

Dollar, John. 2010. "The Man Who Predicted an Earthquake." *Guardian*, April 5, 2010. http://www.theguardian.com/world/2010/apr/05/laquila-earthquake-prediction giampaolo-giuliani.

Domènech, Rossend. 1990. "Political Satire in Italy: A Successful Television Game." In *La satira politica in Italia: Con un'intervista a Tullio Pericoli*, edited by Adolfo Chiesa, 71–78. Bari: Laterza.

Douglas, Mary. 1968. "The Social Control of Cognition: Some Factors in Joke Perception." *Man* 3 (3): 361–376. https://doi.org/10.2307/2798875.

Doyle, Richard. 2003. *Wetwares: Experiments in Postvital Living*. Minneapolis: University of Minnesota Press.

Doyle, Waddick. 2017. "Translating Genres: Translating Leaders: Trump and Berlusconi." *Contemporary French and Francophone Studies* 21 (5): 488–497. https://doi.org/10.1080/17409292.2017.1436198.

Guzzanti, Sabina. 2010. *Draquila: L'Italia Che Trema*. Rome: BIM Distribuzione. DVD video.

Duggan, Christopher. 1984. *A Concise History of Italy*. Cambridge: Cambridge University Press.

Duranti, Alessandro. 2010. "Husserl, Intersubjectivity and Anthropology." *Anthropological Theory* 10 (1): 1–20. https://doi.org/10.1177/1463499610370517.

Dusi, Elena. 2016. "Troppo forte la traccia dall'umo: Benvenuti nell'era dell'antropocene." *La Repubblica*, August 30, 2016. http://www.repubblica.it/ambiente/2016/08/30/news/antropocene_era-146873380/.

Economist. 2004. "It's a Riot." *Economist*, April 3, 2004. https://www.economist.com/node/2553764.

——. 2013. "Italian Banks: Mid-crisis Life." *Economist*, February 23, 2013. https://www.economist.com/finance-and-economics/2013/02/23/mid-crisis-life

Edwards, Paul. 2017. "Knowledge Infrastructures for the Anthropocene." *Anthropocene Review* 4 (1): 34–43. https://doi.org/10.1177/2053019616679854.

Edwards, Phil. 2005. "The Berlusconi Anomaly: Populism and Patrimony in Italy's Long Transition." *South European Society and Politics* 10 (2): 225–243. https://doi.org/10.1080/13608740500134945.

Ekbia, Hamid R., and Bonnie A. Nardi. 2018. "From Form to Content." *Cultural Anthropology* 33 (3): 360–367.

Eleta, Paula. 1997. "The Conquest of Magic over Public Space: Discovering the Face of Popular Magic in Contemporary Society." *Journal of Contemporary Religion* 12 (1): 51–67.

Elzinga, Aant. 2010. "Objectivity and Partisanship in Science." *Ethnos: Journal of Anthropology* 40 (1–4): 406–427. https://www.tandfonline.com/doi/abs/10.1080/00141844.1975.9981114.

Evans-Pritchard, E. E. 1937. *Witchcraft, Oracles, and Magic among the Azande*. Oxford: Clarendon.

Fella, Stefano, and Carlo Ruzza. 2013. "Populism and the Fall of the Centre-Right in Italy: The End of the Berlusconi Model or a New Beginning?" *Journal of Contemporary European Studies* 21 (1): 38–52. https://doi.org/10.1080/14782804.2013.766475.

Fernandez, James W., and Mary Taylor Huber. 2001. "Introduction: The Anthropology of Irony." In *Irony in Action: Anthropology, Practice, and Moral Imagination*, edited by James W. Fernandez and Mary Taylor Huber, 1–31. Chicago: University of Chicago Press.

Ferrari, Chiara, and Michela Ardizzoni. 2010. "Italian Media between the Local and the Global." In *Beyond Monopoly: Globalization and Contemporary Italian Media*, edited by Michela Ardizzoni and Chiara Ferrari, xi–xix. Lanham, MD: Rowman and Littlefield.

Ferraris, Maurizio. 2017. "Perché dobbiamo chiamarla post-verita." *La Repubblica*, May 3, 2017. http://www.libertaegiustizia.it/2017/05/03/perche-dobbiamo-chiamarla-post-verita/.

Fiammeri, Barbara. 2012. "Berlusconi in campo, malumori Pdl." *Il Sole* 24 Ore, July 22, 2012. http://www.ilsole24ore.com/art/notizie/2012-07-22/berlusconi-campo-malumori-081236.shtml?uuid=AbYd3pBG.

Fieschi, Catherine, and Paul Heywood. 2004. "Trust, Cynicism, and Populist Anti-Politics." *Journal of Political Ideologies* 9 (3): 289–309. https://doi.org/10.1080/1356931042000263537.

Fine, Gary Alan. 2007. *Authors of the Storm: Meteorologists and the Culture of Prediction*. Chicago: University of Chicago Press.

Finn, Ed. 2017. *What Algorithms Want: Imagination in the Age of Computing*. Boston: MIT Press.

Fiske, John. 2011 [1987]. *Television Culture*. Abingdon, UK: Routledge.

Flood, Alison. 2016. "'Post-truth' Named Word of the Year by Oxford Dictionaries." *Guardian*, November 15, 2016. https://www.theguardian.com/books/2016/nov/15/post-truth-named-word-of-the-year-by-oxford-dictionaries.

Foucault, Michel. 1972. "Truth and Power." In *Power/Knowledge: Selected and Other Writings 1972–1977*, edited by Colin Gordon, 109–133. New York: Pantheon Books.

——. 2003. *Abnormal: Lectures at Collége de France 1974–1975*. Translated by Graham Burchell, edited by Valerio Marchetti and Antonella Salomoni. New York: Picador.

Francis, His Holiness Pope. 2013. *The Joy of the Gospel: Apostolic Exhortation Evangelii Gaudium of the Holy Father Francis to the Bishops, Clergy, Consecrated Persons and the Lay Faithful on the Proclamation of the Gospel in Today's World*. Frederick, MD: World Among Us Press.

——. 2015. *Laudato Si' of the Holy Father Francis on Care for Our Common Home*. Vatican City: Vatican Press.

——. 2018. "Message of His Holiness Pope Francis for World Communications Day." Vatican, January 18, 2018. https://w2.vatican.va/content/francesco/en/messages/communications/documents/papa-francesco_20180124_messaggio-comunicazioni-sociali.html.

Franklin, Sarah. 2007. *Dolly Mixtures: The Remaking of Genealogy*. Durham, NC: Duke University Press.

Frazer, James. 2009 [1890]. *The Golden Bough: A Study in Magic and Religion*. Oxford: Oxford University Press.

Freud, Sigmund. 1990 [1909]. *Jokes and Their Relation to the Unconscious*. Harmondsworth, UK: Penguin.

Fubini, Federico. 2019. "Walter Ricciardi: 'Lascio l'Istituto superior sanità, il governo ha posizioni antiscientifiche.' *Corriere della Sera*, January 2, 2019. https://www.corriere.it/cronache/19_gennaio_01/difficile-collaborare-il-governo-4a6b4fba-0e01-11e9-991e-8333c5dc4514.shtml.

Fulginiti, Valentina. 2016. "Resisting Leviathan: Depictions of Silvio Berlusconi in Italian Fiction 2003–2011." *Italianist* 36 (1): 106–127. https://doi.org/10.1080/02614340.2015.1120064.

G. R. L. 2011. "Italian Worldviews: Dietrologia." *Economist*, March 11, 2011. https://www.economist.com/blogs/johnson/2011/03/italian_worldviews.

Gervaso, Roberto.1983. *Il grillo parlante.* Milan: Bombiani.

Gilmour, David. 2011. *The Pursuit of Italy: A History of a Land, Its Regions, and Their Peoples.* London: Penguin Group: Allen Lane.

Ginsborg, Paul. 2005. *Silvio Berlusconi: Television, Power and Patrimony.* London: Verso.

Giordano, Cristiana. 2014. *Migrants in Translation: Caring and the Logics of Difference in Contemporary Italy.* Oakland: University of California Press.

Giroux, Henry A. 2010. *Zombie Politics and Culture in the Age of Casino Capitalism.* New York: Peter Lang Publishing.

Giua, Claudio. 2016. "Un algoritmo di nome Rousseau nel destino del M5S di Beppe Grillo." *Il Blitz Quotidiano*, October 1, 2016. https://www.blitzquotidiano.it/opinioni/2558162-2558162/.

Glaeser, Andreas. 2010. *Political Epistemics: The Secret Police, The Opposition, and the End of East German Socialism.* Chicago: University of Chicago Press.

Godfrey-Smith, Peter. 2003. *Theory and Reality: An Introduction to the Philosophy of Science.* Chicago: University of Chicago Press.

Goffredo, Buccini. 1993. "Craxi: Siamo tutti colpevoli." *Corriere della Sera*, December 18, 1993.

Golinski, Jan. 1998. *Making Natural Knowledge: Constructivism and the History of Science.* Cambridge: Cambridge University Press.

Good, Byron. 1994. *Medicine, Rationality, and Experience: An Anthropological Perspective.* Cambridge: Cambridge University Press.

Gordon, Avery. 2008. *Ghostly Matters: Haunting and the Sociological Imagination.* 2nd ed. Minneapolis: University of Minnesota Press.

Graeber, David. 2012. "Can't Stop Believing: Magic and Politics." *Baffler,* no. 21. https://thebaffler.com/salvos/cant-stop-believing.

Graham, Loren. 2013. *Lonely Ideas: Can Russia Compete?* Cambridge, MA: MIT Press.

Grandinetti, Fabio. 2016. "Terremoto, tornano le bufale online: Ecco le notizie a cui non credere." *L'Espresso*, October 27, 2016. http://espresso.repubblica.it/attualita/2016/10/27/news/terremoto-attenti-alle-bufale-1.286719.

Grant, Bruce. 2009. *The Captive and the Gift: Cultural Histories of Sovereignty in Russia and the Caucasus.* Ithaca, NY: Cornell University Press.

Greco, Andrea. 2011. "Antonio Ricci: Ogni italiano ha la sua Wanna Marchi." Striscia la Notizia, November 11. http://www.striscialanotizia.mediaset.it/news/2011/11/30/news_6838.shtml.

Grillo, Beppe. 2012. *Modern Slaves: Precarity in Italy of Wonders.* Casaleggio Associati.

Grosz, Elizabeth. 1996. "Intolerable Ambiguity: Freaks as/at the Limit." In *Freakery: Cultural Spectacles of the Extraordinary Body*, edited by Rosemarie Garland Thomson, 55–66. New York: New York University Press.

Guggenheim, Michael. 2014. "Introduction: Disaster as Political—Politics as Disasters." *Sociological Review* 62 (21): 1–16. https://doi.org/10.1111/1467-954X.12121.

Guinness World Records. 2012. "Longest-Running Satirical News TV Programme." *Guinness World Records*, January. http://community.guinnessworldrecords.com/_Longest-running-satirical-news-TV-programme-by-number-of-episodes/photo/16189473/7691.html.

Gusterson, Hugh. 2019. "Introduction: RoboHumans." In *Life by Algorithms: How RoboProcesses Are Remaking Our World*, edited by Catherine Bestemen and Hugh Gusterson, 1–50. Chicago: University of Chicago Press.

Guyer, Jane. 2007. "Prophecy and the Near Future: Thoughts on Macroeconomic, Evangelical, and Punctuated Time." *American Ethnologist* 34 (3): 409–421. https://doi.org/10.1525/ae.2007.34.3.409.

Haber, Gordon. 2015. "Pope Francis, Science Fiction Lover." *Salon*. March 15, 2015. https://www.salon.com/2015/03/14/pope_francis_science_fiction_lover_partner/.

Habermas, Jurgen. 1984. *The Theory of Communicative Action: Reason and the Rationalization of Society*. Boston: Beacon.

Hacking, Ian. 1983. *Representing and Intervening: Introductory Topics in the Philosophy of Natural Science*. Cambridge: Cambridge University Press.

Hajek, Andrea, and Daniele Salerno. 2014. "Send In the Clowns! Humour and Power in Italian Political, Social and Cultural Life." *Incontri: Rivista Europea di Studi Italiani* 29:3–7. https://doi.org/10.18352/incontri.9873.

Halpern, Michael. 2012. "Italian Scientists Jailed for Failing to Predict Earthquake. Union of Concerned Scientists. http://blog.ucsusa.org/michael-halpern/italian-scientists-jailed-for-failing-to-predict-earthquake.

Hancock, Jeff, Danae Metaxa-Kakavouli, and Joon Park. 2018. "Are Google Search Results Politically Biased?" *Guardian*, September 6, 2018. https://www.theguardian.com/commentisfree/2018/sep/06/google-search-results-rigged-news-donald-trump.

Haraway, Donna. 1988. "Situated Knowledges: The Science Question in Feminism and the 'Privilege of Partial Perspective." *Feminist Studies* 14 (3): 575–599. https://doi.org/10.2307/3178066.

——. 1997. *Modest-Witness@Second-Millenium.Femaleman-Meets-Oncomouse: Feminism and Techoscience*. New York: Routledge.

Harney, Stefano. 2014. "Istituzioni algoritmiche e capitalismo logistico." In *Gli algoritmi del capitale: Accelerazionismo, macchine della conoscenza e autonomia del commune*, edited by Matteo Paquinelli, 116–129. Verona, Italy: Ombre Corte.

Hasian, Marouf, Jr., Nicholas S. Paliewicz, and Robert W. Gehl. 2014. "Earthquake Controversies, the L'Aquila Trials, and the Argumentative Struggles for both Cultural and Scientific Power." *Canadian Journal of Communication* 29:557–576. https://www.cjc-online.ca/index.php/journal/article/view/2740/2488.

Haugerud, Angelique. 2012. "Satire and Dissent in the Age of Billionaires." *Social Research* 79 (1): 145–168. http://www.jstor.org/stable/23350302.

Haugerud, Angelique, Dillon Mahoney, and Meghan Ference. 2012. "Watching *The Daily Show* in Kenya." *Identities: Global Studies in Culture and Power* 19 (2): 168–190. https://doi.org/ 10.1080/1070289X.2012.672853.

Hayden, Cori. 2003. *When Nature Goes Public: The Making and Unmaking of Bioprospecting in Mexico*. Princeton, NJ: Princeton University Press.

Head, Lesley. 2016. *Hope and Grief in the Anthropocene: Re-conceptualising Human-Nature Relations*. London: Routledge.

Helmreich, Stefan. 2009. *Alien Ocean: Anthropological Voyages in Microbial Seas*. Berkeley: University of California Press.

Herzfeld, Michael. 2008. "Mere Symbols." *Anthropologica* 50:141–155. http://www.jstor.org/stable/25605395.

——. 2009. *Evicted from Eternity: The Restructuring of Modern Rome*. Chicago: University of Chicago Press.

Hetherington, Kregg. 2017. "What Came before Post-Truth?" *EASST Review* 36 (2). https://easst.net/article/what-came-before-post-truth/.

Hibberd, Matthew. 2007. "Conflicts of Interest and Media Pluralism in Italian Broadcasting." *West European Politics* 30 (4): 881–902. https://doi.org/10.1080/01402380701500363.

Hochschild, Arlie. 1983. *The Managed Heart: Commercialization of Human Feeling*. Berkeley: University of California Press.

Hooper, John. 2011. "Silvio Berlusconi: Three May Face Trial over 'Bunga Bunga' Parties." *Guardian*, May 6, 2011. http://www.guardian.co.uk/world/2011/may/06/silvio-berlusconi-bunga-bunga-trial.

——. 2013. "Beppe Grillo: Populist Who Could Throw Italy into Turmoil at General Election." *Guardian*, February 11, 2013. https://www.theguardian.com/world/2013/feb/11/beppe-grillo-italy-general-election.

——. 2016. *The Italians*. New York: Penguin Random House Books.

Horowitz, Jason. 2017. "In Italian Schools, Reading, Writing and Recognizing Fake News." *New York Times*, October 18, 2017. https://www.nytimes.com/2017/10/18/world/europe/italy-fake-news.html.

Hufford, David J. 1995. "Beings without Bodies: An Experience-Centered Theory on the Belief in Spirits." In *Out of the Ordinary: Folklore and the Supernatural*, edited by Barbara Walker, 11–45. Logan: Utah State University Press.

"I minerali prodotti dall'uomo raccontano l'Antropocene." 2017. *Le Scienze*, March 2, 2017. http://www.lescienze.it/news/2017/03/02/news/minerali_prodotti_uomo_antropocene-3444623/.

"Il regno degli Algoritmi." 2018. *Il Blog di Beppe Grillo*, June 3, 2018. http://www.beppegrillo.it/il-regno-degli-algoritmi-saremo-costretti-a-credere-alle-macchine/

"Illeciti sulla privacy dei voti in Rousseau." 2018. *La repubblica*, January 2, 2018. https://www.repubblica.it/politica/2018/01/02/news/_illeciti_sulla_privacy_dei_dati_personali_in_rousseau_il_garante_bacchetta_il_m5s-185681865/.

"In Quotes: Berlusconi in His Own Words." 2006. BBC, May 2, 2006. http://news.bbc.co.uk/2/hi/europe/3041288.stm.

Innocenzi, Giulia. 2017. "'Vacci-nazione'—Ecco un estratto del libro di Giulia Innocenzi." *Il Fatto Quotidiano*, December 7, 2017. https://www.ilfattoquotidiano.it/premium/articoli/prima-lavora-con-le-aziende-poi-scrive-la-legge-sui-vaccini/.

"Internet Users in the World." 2017. *Internet World Stats*. https://www.internetworldstats.com/stats.htm.

"Intervista RTS—8 Febbraio 2018." 2018. *Blog di Beppe Grillo*, February 12, 2018. http://www.beppegrillo.it/intervista-rts-8-febbraio-2018/.

Io C'é. 2018. Directed by Alessandro Aronadio, performances by Massimo Alberti, Adriana Alberti, and Giuseppe Battiston. Vision Distribution.

Iovino, Serenella. 2016. *Ecocriticism and Italy: Ecology, Resistance, and Liberation*. London: Bloomsbury Academic.

Islekel, Ege Selin. 2016. "Ubu-esque Sovereign, Monstrous Individual: Death in Biopolitics." *Philosophy Today* 60 (1): 175–191.

"Italian Elections 2018—Full Results." 2018. *Guardian*, March 5, 2018. https://www.theguardian.com/world/ng-interactive/2018/mar/05/italian-elections-2018-full-results-renzi-berlusconi.

Jackson, Jasper. 2017. "Eli Pariser: Activist Whose Filter Bubble Warnings Presages Trump and Brexit." *Guardian*, January 8, 2017. https://www.theguardian.com/media/2017/jan/08/eli-pariser-activist-whose-filter-bubble-warnings-presaged-trump-and-brexit.

Jackson, Michael. 2005. *Existential Anthropology: Events, Exigencies, and Effects*. Berghahn Books.

Jameson, Fredric. 1991. *Postmodernism, Or the Cultural Logic of Late Capitalism*. Durham, NC: Duke University Press.

——. 1995. *The Geopolitical Aesthetic: Cinema and Space in the World System*. Bloomington: Indiana University Press.

Jebreal, Rula. 2015. "Donald Trump Is America's Silvio Berlusconi." *Washington Post*, September 21, 2015. https://www.washingtonpost.com/posteverything/wp/2015/09/21/donald-trump-is-americas-silvio-berlusconi/?noredirect=on&utm_term=.e48a52be8c68.

Jones, Erik. 2009. "Wheeler Dealers: Silvio Berlusconi in Comparative Perspective." *Journal of Modern Italian Studies* 14 (1): 38–45. https://doi.org/10.1080/13545710802642875.

Jones, Graham M. 2010. "Modern Magic and the War on Miracles in French Colonial Culture." *Comparative Studies in Society and History* 52 (1): 66–99. http://www.jstor.org/stable/40603072.

Jules-Rosette, Bennetta. 1978. "The Veil of Objectivity: Prophecy, Divination, and Social Inquiry." *American Anthropologist* 80 (3): 549–570. https://doi.org/10.1525/aa.1978.80.3.02a00020.

Kapferer, Bruce 2007. "Anthropology. The Paradox of the Secular." *Social Anthropology/Anthropologies Sociale* 9 (3): 341–344.

Kendzior, Sarah. 2011. "Digital Distrust: Uzbek Cynicism and Solidarity in the Internet Age." *American Ethnologist* 38 (3): 559–575. https://doi.org/10.1111/j.1548-1425.2011.01323.x.

Kenesson, Frank G. 1993. "Statecraft of the Absurd." *History of European Ideas* 17 (4): 427–437.

Kennedy, Paul. 2017. *Vampire Capitalism: Fractured Societies and Alternative Futures*. London: Palgrave Macmillan UK.

Kertzer, David. 2014. *The Pope and Mussolini: The Secret History of Pius XI and the Rise of Fascism in Europe*. New York: Random House.

——. 2002. *The Pope against the Jews: The Vatican's Role in the Rise of Modern Anti-Semitism*. New York: Alfred A. Knopf.

Killinger, Charles L. 1992. *The History of Italy*. Westport, CT: Greenwood.

Kirchgaessner, Stephanie. 2016. "If Berlusconi Is Like Trump, What Can America Learn from Italy?" *Guardian,* Nov 21, 2016. https://www.theguardian.com/world/2016/nov/21/if-berlusconi-is-like-trump-what-can-italy-teach-america.

Klein, Naomi. 2007. *The Shock Doctrine: The Rise of Disaster Capitalism*. New York: Picador.

Klein, Roberta, and Roger A. Pielke Jr. 2002. "Bad Weather? Then Sue the Weatherman! Part 1: Legal Liability for Public Sector Forecasts." *American Meteorological Society BAMS* 83 (12): 1791–1799. https://doi.org/10.1175/BAMS-83-12-1791.

Klinenberg, Eric. 2013. *Going Solo: The Extraordinary Rise and Surprising Appeal of Living Alone*. New York: Penguin Books.

Klumbytė, Neringa. 2014. "Of Power and Laughter: Carnivalesque in Politics and Moral Citizenship in Lithuania." *American Ethnologist* 41 (3): 473–490. https://doi.org/10.1111/amet.12088.

Koertge, Noretta, ed. 1998. *A House Built on Sand: Exposing Postmodernist Myths about Science*. Oxford: Oxford University Press.

Kohn, Eduardo. 2014. "Toward an Ethical Practice in the Anthropocene." *HAU: Journal of Ethnographic Theory* 4 (1): 459–464. https://doi.org/10.14318/hau4.1.028.

Kolbert, Elizabeth. 2015. "The Shaky Science behind Predicting Earthquakes." *Smithsonian Magazine*. http://www.smithsonianmag.com/science-nature/shaky-science-behind-predicting-earthquakes-180955296/?no-ist.

Koopman, Colin. 2015. "The Algorithm and the Watchtower." *New Inquiry*, September 29, 2015. http://thenewinquiry.com/essays/the-algorithm-and-the-watchtower/.

——. 2019. *How We Became Our Data: A Genealogy of the Informational Person*. Chicago: University of Chicago Press.

Kramer, Jane. 2015. "Demolition Man." *New Yorker*, June 29, 2015, 36–47. https://www.newyorker.com/magazine/2015/06/29/the-demolition-man.

Latella, Maria. 2000. "Nani e ballerina: La corte dei miracoli nata e cresciuta intorno a Bettino." *Corriere della Sera*, January 8, 2000.

Latour, Bruno. 1979. *Laboratory Life: The Social Construction of Facts*. Princeton, NJ: Princeton University Press.

——. 1987. *Science in Action: How to Follow Scientists and Engineers through Society*. Cambridge, MA: Harvard University Press.

——. 1993. *We Have Never Been Modern*. Cambridge, MA: Harvard University Press.

——. 1999. *Pandora's Hope: Essays on the Reality of Science Studies*. Cambridge, MA: Harvard University Press.

——. 2004. "Why Has Critique Run out of Steam? From Matters of Fact to Matters of Concern." *Critical Inquiry* 30:225–248. https://doi.org/10.1086/421123.

——. 2010. *On the Modern Cult of the Factish Gods*. Durham, NC: Duke University Press.

——. 2014. "Agency at the Time of the Anthropocene." *New Literary History* 45 (1) (winter): 1–18. https://doi.org/10.1353/nlh.2014.0003.

Latour, Bruno, and Steve Woolgar. 1986. *Laboratory Life: The Construction of Scientific Facts*. Princeton, NJ: Princeton University Press.

Laveaga, Gabriela Soto. 2009. *Jungle Laboratories: Mexican Peasants, National Projects, and the Making of the Pill*. Durham, NC: Duke University Press.

Leavitt, John. 2000. "Prophecy." *Journal of Linguistic Anthropology* 9 (1–2): 201–204. https://doi.org/10.1525/jlin.1999.9.1-2.201.

Leiss, William, Stephen Kline, Sut Jhally, Jackie Botteril, and Kyle Asquith. 2018. *Social Communication in Advertising, Fourth Edition*. New York: Routledge.

Lewis, Paul. 2006. *Cracking Up: American Humor in a Time of Conflict*. Chicago: University of Chicago Press.

Loucaides, Darren. 2019. "Careful What You Wish For: The Inside Story of Italy's Five Star Movement and the Cyberguru Who Dreamed It Up." *Wired*, March 2019, 80–95.

Lowrie, Ian. 2018. "Algorithms and Automation: An Introduction." *Cultural Anthropology* 33 (3): 349–359. http://doi.org/10.14506/ca33.3.01.

Luhrmann, T. M. 2012. *When God Talks Back: Understanding the American Evangelical Relationship with God*. New York: Vintage Books, Random House.

Lukacs, Gabriella. 2010. *Scripted Affects, Branded Selves: Television, Subjectivity, and Capitalism in 1990s Japan*. Durham, NC: Duke University Press.

Lukes, H. N. 2016. "Perversion, Terminable and Interminable: Foucault, Lacan and *DSM-5*." *Discourse* 38 (3): 327–355.

Lupton, Deborah. 2015. *Digital Sociology*. London: Routledge.

Lyotard, Jean-François. 1984. *The Postmodern Condition: A Report on Knowledge*. Translated by G. Bennington and Brian Massumi. Minneapolis: University of Minnesota Press.

Mackenzie, Adrian. 2015. "The Production of Prediction: What Does Machine Learning Want?" European Journal of Cultural Studies 18 (4–5): 429–445. https://doi.org/10.1177/1367549415577384.

Madrigal, Alexis C. 2018. "No, You Don't Really Look Like That." *Atlantic*, December 18, 2018. https://www.theatlantic.com/technology/archive/2018/12/your-iphone-selfies-dont-look-like-your-face/578353/?utm_source=facebook&utm_campaign=the-atlantic&utm_content=edit-promo&utm_term=2018-12-18T20%3A54%3A27&utm_medium=social.

Magliocco, Sabina. 2004. "Witchcraft, Healing and Vernacular Magic in Italy." In *Witchcraft Continued: Popular Magic in Modern Europe*, edited by Willem de Blecourt and Owen Davies, 151–173. Manchester, UK: Manchester University Press.

——. 2012. "Beyond Belief: Context, Rationality and Participatory Consciousness." *Western Folklore* 71 (winter): 5–24. https://www.jstor.org/stable/24550769.

Mahmud, Lilith. 2014. *The Brotherhood of Freemason Sisters: Gender, Secrecy, and Fraternity in Italian Masonic Lodges*. Chicago: University of Chicago Press.

Malinowski, Bronislaw. 2015 [1954]. *Magic, Science and Religion*. Eastford, CT: Martino Fine Books.

Mankekar, Purnima. 1999. *Screening Culture, Viewing Politics: An Ethnography of Television, Womanhood, and Nation in Postcolonial India*. Durham, NC: Duke University Press.

Mapei, Camillo. 1864. "The Political, Religious, and Moral State of Italy." In *Italy, Illustrated and Described with a Review of Its Past Condition and Future Prospects,* edited by Carlyle Gavin and Camillo Mapei. Translated by David Dundes Scott. London: Blackie and Son.

Marchisio, Roberto, and Maurizio Pisati. 1999. "Belonging without Believing: Catholics in Contemporary Italy." *Journal of Modern Italian Studies* 4 (2): 236–255. https://doi.org/10.1080/13545719908455008.

Marco, Imarisio. 2009. "E la Cassandra attaca: Bugiardi, dovevano ascoltare." *Corriere della Sera*, April 7, 2009. http://www.poloserviziculturaliabruzzo.it/sismaq/rassegnastampa/corrieredellasera/pdf/Corrieredellasera_2009040704.pdf.

Marcus, George, 2000. "Introduction." *Para-Sites: A Casebook against Cynical Reason*. Edited by George Marcus, 1–14. Chicago: University of Chicago Press.

Marshall, Jonathan Paul, James Goodman, Didar Zowghi, and Francesca da Rimini. 2015. *Disorder and the Disinformation Society: The Social Dynamics of Information, Networks, Software*. New York: Routledge.

Masco, Joseph. 2006. *Nuclear Borderlands: The Manhattan project in Post-Cold War New Mexico*. Princeton, NJ: Princeton University Press.

——. 2014. *The Theater of Operations: National Security Affect from the Cold War to the War on Terror*. Durham, NC: Duke University Press.

Massumi, Brian. 2011. *Semblance and Event: Activist Philosophy and the Occurrent Arts*. Cambridge, MA: MIT Press.

Mathews, Andrew S. 2018. "Landscapes and Throughscapes in Italian Forest Worlds: Thinking Dramatically about the Anthropocene." *Cultural Anthropology* 33 (3): 386–414. https://doi.org/10.14506/ca33.3.05.

Matteucci, Piera. 2012. "Terremoto L'Aquila, scienzati divisi sulla sentenza contro la Grandi rischi." *La Repubblica*, October 23, 2012. http://www.repubblica.it/cronaca/2012/10/23/news/l_aquila_scienziati_usa_contro_condanna-45125743/.

Mauri, Paolo. 2012. "Terremoti e prevedibilita': L'Aquila e il caso Giuliani." *Il Primato Nazionale*, January 16, 2012. http://www.ilprimatonazionale.it/cronaca/terremoti-e-prevedibilita-laquila-e-il-caso-giuliani-4725/.

Mautino, Beatrice. 2012. "Beppe Grillo e il negazionismo: Un bignamino sulle caratteristiche delle teorie del complotto." *Wired*, May 23, 2012. http://blog.wired.it/barnum/2012/05/23/beppe-grillo-e-il-negazionismo.html.

Mbembe, Achille. 1992. "Provisional Notes on the Postcolony." *Africa* 62 (1): 3–37. https://doi.org/10.2307/1160062.

Mbembe, Achille, and Janet Roitman. 1995. "Figures of the Subject in Times of Crisis." *Public Culture* 7:323–352. https://doi.org/10.1215/08992363-7-2-323.

McClennen, Sophia A. 2016. "Colbert Goes after Trumpiness." *Salon*, July 22, 2016. https://www.salon.com/2016/07/22/colbert_goes_after_trumpiness_his_live_rnc_coverage_revives_the_comedy_of_the_colbert_report/.

McGuire, Bill. 2012. *Waking the Giant: How a Changing Climate Triggers Earthquakes, Tsunamis, and Volcanoes*. New York: Oxford University Press.

McNally, David. 2011. *Monsters of the Market: Zombies, Vampires and Global Capitalism*. Leiden, The Netherlands: Koninklijke Brill NV.

McQuillan, Don. 2015. "Algorithmic States of Exception." *European Journal of Cultural Studies* 18 (4–5): 564–576. https://doi.org/10.1177/1367549415577389.

Melanco, Mirco. 1995. "Italian Cinema, since 1945: The Social Costs of Industrialization; Historical." *Journal of Film, Radio and Television* 15 (3): 387–392. https://doi.org/10.1080/01439689500260261.

Merlo, Francesco. 2011. "Da porta a porta al bunga bunga." *La Repubblica*, October 9, 2011. http://www.repubblica.it/politica/2011/10/09/news/berlusconi-22924782/index.html?ref=search.

Messina, Sebastiano. 2002. "Aneddoti, gaffe e amici illustri: L'anno magico di Silvio in felucca." *La Repubblica*, November 14, 2002. http://www.repubblica.it/online/politica/bossitrem/feluche/feluche.html?ref=search.

Messora, Claudio. 2009. "Il cacciatore di terremoti," *Byoblu*, April 15, 2009. http://www.byoblu.com/post/2009/04/15/Il-cacciatore-di-terremoti.aspx.

"M5s, ecco 'Rousseau.'" 2015. *La Repubblica*, July 17, 2015. https://www.repubblica.it/tecnologia/mobile/2015/07/17/news/m5s_ecco_rousseau_sistema_operativo_per_il_movimento_anche_mobile-119292454/.

"M5s, il simbolo del Movimento concesso in uso gratuito all'Associazione Rousseau." 2018. *Il Fatto Quotidiano*, October 7, 2018. https://www.ilfattoquotidiano.it/2018/10/07/m5s-il-simbolo-del-movimento-concesso-in-uso-gratuito-allassociazione-rousseau/4676277/.

Michaels, Adrien. 2007. "Naked Ambition." *Financial Times* (London), July 14, 2007. https://www.ft.com/content/7d479772-2f56-11dc-b9b7-0000779fd2ac.

Mignone, Mario B. 2008. *Italy Today: Facing the Challenges of the New Millennium*. New York: Peter Lang.

Minuz, Andrea. 2018. "L'algoritmo Di Maio." *Il Foglio*, February 18, 2018. https://www.ilfoglio.it/politica/2018/02/19/news/m5s-di-maio-maschera-del-grillismo-179559/.

Mishra, Pankaj. 2016. "How Rousseau Predicted Trump." *New Yorker*, August 1, 2016. https://www.newyorker.com/magazine/2016/08/01/how-rousseau-predicted-trump.

Mittelstrass, Jurgen. 2010. "The Loss of Knowledge in the Information Age." In *From Information to Knowledge, From Knowledge to Wisdom: Challenges and Changes Facing Higher Education in the Digital Age*, edited by Erik De Corte, 19–23. London: Portland.

Moe, Nelson. 2002. *The View from Vesuvius: Italian Culture and the Southern Question*. Berkeley: University of California Press.

Mohammed, Shaheed Nick. 2012. *(Dis)information Age: The Persistence of Ignorance*. New York: Peter Lang.

Molchan, G., and L. Romashkova. 2013. "Earthquake Prediction Analysis: The M8 Algorithm." *Arxiv.org*. https://arxiv.org/abs/1307.3464v1.

Molé, Noelle. 2011. *Labor Disorders in Neoliberal Italy: Mobbing, Well-being and the Workplace*. Bloomington: Indiana University Press.

——. 2013a. "Trusted Puppets, Tarnished Politicians: Humor and Cynicism in Berlusconi's Italy." *American Ethnologist* 40 (2): 288–299. https://doi:10.1111/amet12021.

——. 2013b. "Enchanting the Disenchanted: Grillo's Supernatural Humor as Populist Politics." *Perspectives on Europe* 43 (2): 7–14.

Mosca, Lorenzo, and Filippo Tronconi. 2019. "Beyond Left and Right: The Eclectic Populism of the Five Star Movement. *West European Politics* 42 (6): 1258–1283. https://doi.org/ 10.1080/01402382.2019.1596691.

Moseman, Andrew. 2009. "Does Cloud Seeding Work?" *Scientific American,* February 19, 2009. https://www.scientificamerican.com/article/cloud-seeding-china-snow/.

Muehlebach, Andrea. 2012. *The Moral Neoliberal: Welfare and Citizenship in Italy*. Chicago: University of Chicago Press.

Murphy, Keith M. 2015. *Swedish Design: An Ethnography*. Ithaca, NY: Cornell University Press.

Nafus, Dawn. 2018. "Exploration or Algorithm? The Undone Science Before the Algorithms." *Cultural Anthropology* 33 (3): 368–374.

Natale, Simone, and Andrea Ballatore. 2014. "The Web Will Kill Them All: New Media, Digital Utopia, and Political Struggle in the Italian 5-Star Movement." *Media Culture & Society* 36 (1): 105–121. https://doi.org/10.1177/0163443713511902.

Navarria, Giovanni. 2019. *The Networked Citizen: Power, Politics, and Resistance in Internet Age*. Singapore: Palgrave Macmillan.

Newitz, Annalee. 2006. *Pretend We're Dead: Capitalist Monsters in American Pop Culture*. Durham, NC: Duke University Press.

Newman, Lorenzo. 2017. "Bunga Bunga, American Style." *Slate*, April 3, 2017. http://www.slate.com/articles/news_and_politics/foreigners/2017/04/the_similarities_between_trump_and_berlusconi_are_much_deeper_than_you_think.html.

Nicodemo, Francesco. 2018. *Disinformazia: La Comunicazione al tempo dei social media*. Padova, Italy: Marsilio.

Nisbet, Matt. 2000. "Introducing Italy's Version of Harry Houdini." *Skeptical Inquirer* 25 (2): 43–45. https://www.csicop.org/specialarticles/show/introducing_italys_version_of_harry_houdini.

Novella, Steven, and David Bloomberg. 1999. "Scientific Skepticism, CSICOP, and the Local Groups." *Skeptical Inquirer* 23 (4): 44–46. https://www.csicop.org/si/show/scientific_skepticism_csicop_and_the_local_groups.

Nuckolls, Charles W. 1991. "Culture and Causal Thinking: Diagnosis and Prediction in a South Indian Fishing Village." *Ethos* 19 (1): 3–51. https://doi.org/10.1525/eth.1991.19.1.02a00010.

"Number of Mobile Phone Users Worldwide from 2013 to 2019." 2018. *Stastica*. https://www.statista.com/statistics/274774/forecast-of-mobile-phone-users-worldwide/.

O'Leary, Alan. 2010. "Italian Cinema and the 'Anni di Piombo.'" *Journal of European Studies* 40 (3): 243–257. https://doi.org/10.1177/0047244110371912.

O'Neil, Cathy. 2016. *Weapons of Math Destruction: How Big Data Increases Inequality and Threatens Democracy*. London: Penguin.

"Onore alla Democrazia." 2018. *Blog di Beppe Grillo*, February 9, 2018. http://www.beppegrillo.it/onore-alla-democrazia/.

"OPERA News and Updates." 2018. OPERA Web, May 22. http://operaweb.lngs.infn.it/spip.php?rubrique14.

Oremus, Will. 2013. "Google Knows What You're Doing Tomorrow." *Slate*, August 14, 2013. http://www.slate.com/blogs/future_tense/2013/08/14/google_search_gets_scary_smart_with_flight_reservation_photo_search_features.html.

Orr, Jackie. 2006. *Panic Diaries: A Genealogy of Panic Disorder*. Durham, NC: Duke University Press.

Orsi, Robert A. 2007. "When 2 + 2=5." *American Scholar*, March 1, 2007. https://theamericanscholar.org/when-2-2-5/.

Orsina, Giovanni. 2014. *Berlusconism and Italy: A Historical Interpretation*. New York: Palgrave Macmillan.

Panarari, Massimiliano. 2010. *L'egemonia sottoculturale: L'Italia da Gramsci al gossip*. Turin: Einaudi.

Para. 2017. *The American Heritage Science Dictionary*. Dictionary.com. http://www.dictionary.com/browse/para-.

Pariser, Eli. 2011. *The Filter Bubble: What the Internet Is Hiding from You*. London: Penguin/Viking.

Parsons, Elsie Clews. 1942. "Anthropology and Prediction." *American Anthropologist* 44 (3): 337–344. https://doi.org/10.1525/aa.1942.44.3.02a00010.

Pasquale, Frank. 2015. "The Algorithmic Self." *Hedgehog Review* Spring: 30–45. http://www.iasc-culture.org/THR/THR_article_2015_Spring_Pasquale.php.

Pasquinelli, Matteo, "Introduzione." 2014a. In *Gli algoritmi del capital: Accelerazionismo, machine della conoscenza e autonomia del commune*, edited by Matteo Pasquinelli, 7–16. Verona, Italy: Ombre Corte.

Paumgarten, Nick. 2014. "Make Me a Match." *New Yorker*, August 25, 2014. https://www.newyorker.com/magazine/2014/08/25/2710913.

Perasso, Eva. 2015. "Antropocene: L'attuale era geologica è inziata con la prima bomba atomica." *Corriere della Sera*, January 23, 2015. http://www.corriere.it/scienze/15_gennaio_21/antropocene-era-geologica-bomba-atomica-aff44af6-a188-11e4-8f86-063e3fa7313b.shtml.

Pietrucci, Pamela, and Leah Ceccarelli. 2019. "Scientist Citizens: Rhetoric and Responsibility in L'Aquila." *Rhetoric & Public Affairs* 22 (1): 95–128. https://muse.jhu.edu/article/718702.

Pinch, Trevor, and Wiebe Bijker. 1987. "The Social Construction of Facts and Artifacts." In *The Social Construction of Technological Systems*, edited by Wiebe E. Bijker, Thomas P. Hughes, and Trevor J. Pinch, 17–51. Cambridge, MA: MIT Press.

Pine, Jason. 2012. *The Art of Making Do in Naples*. Minneapolis: University of Minnesota Press.

Pirandello, Luigi. 1974. *On Humor*. Translated by Antonio Illiano and Daniel P. Testa. Chapel Hill: University of North Carolina Press.

Pisa, Nick. 2010. "Italy's Culture Minister in Cannes Film Festival Boycott." *Telegraph*, May 10, 2010. https://www.telegraph.co.uk/news/worldnews/europe/italy/7701590/Italys-Culture-Minister-in-Cannes-Film-Festival-boycott.html.

Polidoro, Massimo. 2008. "Hunting for Spooklights." *Skeptical Inquirer* 32 (5), September/October: 27–29. https://www.csicop.org/si/show/hunting_for_spooklights.

———. 2011. "Tutti alla III Giornata Anti-Superstizione." *Cicap.org*, May 5, 2011. https://www.cicap.org/n/articolo.php?id=274509.

Poovey, Mary. 1998. *A History of the Modern Fact: Problems of Knowledge in the Sciences of Wealth and Society*. Chicago: University of Chicago Press.

Prakash, Gyan. 1999. *Another Reason: Science and the Imagination of Modern India*. Princeton, NJ: Princeton University Press.

"Presentato Il Libro Realizzato dall'Osservatorio dell'Universita." *Abbruzzo Web*. http://www.abruzzoweb.it/contenuti/l-aquila-il-terremoto-tra-norme-ignorate-e-contraddizioni/29401-302/.

Pucciarelli, Matteo. 2018. "M5S, un algoritmo per il governo: Ecco il prof di Di Maio che valuta l'affinità con Lega e Pd." *La Repubblica*, April 13, 2018. https://www.repubblica.it/politica/2018/04/13/news/m5s_un_algoritmo_per_il_governo_ecco_il_team_di_di_maio_che_valuta_l_affinita_con_lega_e_pd-193731387/.

Purdy, Jedediah. 2015. *After Nature: A Politics for the Anthropocene*. Cambridge, MA: Harvard University Press.

Quack, Johannes. 2012. *Disenchanting India: Organized Rationalism and Criticism of Religion in India*. Oxford: Oxford University Press.

Quattrociocchi, Walter, and Antonella Vicini. 2016. *Misinformation: Guida alla società dell'informazione e della credulità*. Milan: Franco Angeli.

Rabin, Nathan. 2006. "Stephen Colbert." *The AVClub*, January 25, 2006. http://www.avclub.com/article/stephen-colbert-13970.

Rabinow, Paul. 1999. *French DNA: Trouble in Purgatory*. Chicago: University of Chicago University.

Raffnsøe, Sverre. 2016. *Philosophy of the Anthropocene: The Human Turn*. Basingstoke: Palgrave Macmillan.

Ramage, Craufurd Tait. 1868. *Italy, Wanderings: In Search of Its Ancient Remains and Modern Superstitions*. Liverpool: Edward Howell.

"Renzi: M5S Partito-Algoritmo." 2017. *Vista Agenzia Televisiva Nazionale*, May 21, 2017. YouTube. https://www.youtube.com/watch?v=IVqfoVX8xyI.

Ricci, Antonio. 1998. *Striscia la Tivú: Comicità spettacolo informazione*. Turin: Giulio Einaudi Rizzini.

Roberts, Tobias. 2017. "Do We Have a Way Out of the Anthropocene?" *Huffington Post*, August 15, 2017. http://www.huffingtonpost.com/entry/do-we-have-a-way-out-of-the-anthropocene_us_59931c74e4b0afd94eb3f521.

Rosenberg, Matthew, Nicholas Confessore, and Carole Cadwalladr. 2018. "How Trump Consultants Exploited the Facebook Data of Millions." *New York Times*, March 17, 2018. https://www.nytimes.com/2018/03/17/us/politics/cambridge-analytica-trump-campaign.html.

Rossiter, Ned, and Soenke Zehle. 2014. "Experience Machines." *Sociologia del Lavoro* 133:111–133. doi:10.3280/SL2014-133008.

Rotondi, Armando. 2017. "Communication and 'Theatralization' of the Italian Crisis in the Dialectic between Dario Fo, Beppe Grillo and Gianroberto Casaleggio." *Journal of Media Critiques* 3 (10): 194–203. doi:10.17349/jmc117212.

Roudakova, Natalia. 2017. *Losing Pravda: Ethics and The Press in Post-Truth Russia*. Cambridge: Cambridge University Press.

Rudlin, John. 1994. *Commedia dell'Arte: An Actor's Handbook*. New York: Routledge.

Salvadorini, Ranieri. 2013. "L'Aquila: processo alla scienza o alla negligenza?" June 16, 2013. https://www.scienzainrete.it/contenuto/articolo/ranieri-salvadorini/laquila-processo-alla-scienza-o-alla-negligenza/giugno-2013.

Salvaggiulo, Giuseppe. 2011. "La Bocconi Scopre il valore di 'Striscia la Notizia.'" *La Stampa*, November 29, 2011. http://www3.lastampa.it/spettacoli/sezioni/articolo/lstp/432230/.

Sample, Ian. 2009. "Scientist Was Told to Remove Internet Prediction of Italy Earthquake." *Guardian*, April 6, 2009. https://www.theguardian.com/world/2009/apr/06/italy-earthquake-predicted.

Sampson, Tim. 2014. "Researchers Can Now Predict Likelihood of Underage Binge Drinking." *Daily Dot*, July 3, 2014. https://www.dailydot.com/debug/computer-model-predicts-underage-binge-drinking/.

Sanburn, Josh. 2011. "Scientists Create 52 Artificial Rain Storms in Abu Dhabi Desert." *Time*, January 3, 2011. https://newsfeed.time.com/2011/01/03/scientists-create-52-artificial-rain-storms-in-abu-dhabi-desert/.

Sargiacomo, Massimo. 2015. "Earthquakes, Exceptional Government and Extraordinary Accounting." *Accounting, Organizations and Society* 42:67–89. doi:10.1016/i.aos.2015.02.001.

Sayre, Nathan F. 2010. "The Politics of the Anthropogenic." *Annual Review of Anthropology* 41:57–70. https://doi.org/10.1146/annurev-anthro-092611-145846.

Scanzi, Andrea. 2012. *Ve lo do io Beppe Grillo*. Milan, Italy: Mondadori.

Scheper-Hughes, Nancy. 1996. "Theft of Life: The Globalization of Organ Stealing Rumours." *Anthropology Today* 12 (3): 3–11.

"Scie Chimiche." Scie Chimiche.org. Primo sito italiano sulle scie chimiche.

Schüll, Natasha. 2016. "Data for Life: Wearable Technology and the Design of Self-Care." *BioSocieties* (October):1–17. https://doi: 10.1057/biosoc.2015.47.

Seaver, Nick. 2018. "What Should an Anthropology of Algorithms Do?" *Cultural Anthropology* 33 (3): 375–385.

"Sentenza L'Aquila/Giuliani: il terremoto l'ho previsto, ma quel giorno non mi fecero salvare delle vite." 2012. *Il Sussidiario*.net, October 23. http://www.ilsussidiario.net/News/Cronaca/2012/10/23/SENTENZA-L-AQUILA-Giuliani-il-terremoto-l-ho-previsto-ma-quel-giorno-non-mi-fecero-salvare-delle-vite/2/331451.

Serres, Michel. 1982. *The Parasite*. Baltimore: John Hopkins University Press.

Severgnini, Beppe. 2011. *Mamma Mia: Berlusconi Explained for Posterity and Friends Abroad*. Milan: Rizzoli ex libris.

——. 2016. "What a Trump America Can Learn from Berlusconi Italy," *New York Times*, November 15, 2016. https://www.nytimes.com/2016/11/16/opinion/what-a-trump-america-can-learn-from-a-berlusconi-italy.html.

Shanafelt, Robert. 2004. "Magic, Miracle, and Marvels in Anthropology." *Ethnos: Journal of Anthropology* 69 (3): 317–340. https://doi.org/10.1080/0014184042000260017.

Shapin, Steven. 1994. *A Social History of Truth: Civility and Science in Seventeenth-Century England*. Chicago: University of Chicago Press.

——. 2014. *The Scientific Life: A Moral History of a Late Modern Vocation*. Chicago: University of Chicago Press.

Shin, Michael E., and John A. Agnew. 2008. *Berlusconi's Italy: Mapping Contemporary Italian Politics*. Philadelphia: Temple University Press.

Simione, Corrado, ed. 1970. *Il teatro di Luigi Pirandello: L'innesto, La patente, L'uomo, La bestia e la virtu*. Rome: Oscar Mondadori.

Singer, Natasha. 2014. "Today's Students Don't Have to Suffer If They Hate Their Roommates." *New York Times*, July 20, 2014. https://bits.blogs.nytimes.com/2014/07/20/todays-students-dont-have-to-suffer-if-they-hate-their-roommates/.

"Sisma L'L'Aquila, sentenza Grandi Rischi: 'Affermazioni approssimative e inefficacy.'" 2013. *Il Gazettino*, January 18, 2013. http://www.gazzettino.it/italia/cronacanera/sisma_L'Aquila_sentenza_grandi_rischi_affermazioni_approssimative_e_inefficaci/notizie/245162.shtml.

Sismondo, Sergio. 1996. *Science without Myth: On Constructions, Reality, and Social Knowledge*. SUNY Series in Science, Technology, and Society. Edited by Sal Restivo and Jennifer Croissant. Albany: SUNY Press.

Sluhovsky, Moshe. 2007. *Believe Not Every Spirit: Possession, Mysticism & Discernment in Early Modern Catholicism*. Chicago: University of Chicago Press.

Smith, Chris, and Ben Voth. 2002. "The Role of Humor in Political Argument: How 'Strategery' and 'Lockboxes' Changed a Political Campaign." *Argumentation and Advocacy* 39:110–129. https://doi.org/10.1080/00028533.2002.11821580.

Solla, Gianluca. 2011. "L'Osceno: La Societa' immaginaria e la fine dell'esperienza." In *Filosofia di Berlusconi: L'essere e il nulla nell'Italian del Cavaliere*, edited by Carlo Chiurco, 129–161. Verona: Ombra Corte.

Striphas, Ted. 2015. "Algorithmic Culture." *European Journal of Cultural Studies* 18 (4–5): 395–412. https://doi.org/10.1177/1367549415577392.

Striscia la Notizia. 2010a. "La foto di gruppo." May 31, 2010. http://www.striscialanotizia.mediaset.it/video/videoflv.shtml?2010_06_poli4.flv.

——. 2010b. "Un improvviso abbiocco." February 8, 2010. http://www.striscialanotizia.mediaset.it/video/videoflv.shtml?2010_02_poli8.flv.

——. 2011a. "A rischio licenziamento." February 26, 2011. http://www.striscialanotizia.mediaset.it/video/videoextra.shtml?12744.

——. 2011b. "Politicanti." January 17, 2011. http://www.striscialanotizia.mediaset.it/video/videoflv.shtml?2011_01_musi17.flv.

——. 2011c. "Tapirone a Berluscone." November 9, 2011. http://www.striscialanotizia.mediaset.it/video/videoextra.shtml?13948.

——. 2011d. "Un tapiro bunga-bunga." October 29, 2011. http://www.striscialanotizia.mediaset.it/video/videoflv.shtml?2010_10_tapi29.flv.

——. 2012a. "Gabibbo videogallery." April 6, 2012. http://www.striscialanotizia.mediaset.it/videogallery/videogallery_gabibbo.shtml.

——. 2012b. "Premio a Striscia la Notizia." June 1, 2012. http://www.striscialanotizia.mediaset.it/template/template_premistriscia06_10.shtml.

Sunstein, Cass R. 2017. *#Republic: Divided Democracy in the Age of Social Media*. Princeton, NJ: Princeton University Press.

Tambiah, Stanley J. 1990. *Magic, Science, Religion and the Scope of Rationality*. Cambridge: Cambridge University Press.

Tanzarella, Mariella. 1988. "Striscia la Notizia:1988 Guardateci con 'Odiens.'" *La Repubblica*, November 29, 1988. http://ricerca.repubblica.it/repubblica/archivio/repubblica/1988/11/30/guardateci-con-odiens.html.

Taussig, Karen-Sue. 2009. *Ordinary Genomes: Science, Citizenship, and Genetic Identities*. Durham, NC: Duke University Press.

Taylor, Adam. 2016. "Is Trump a Berlusconi? Let a Berlusconi Expert Explain." *Washington Post*, November 16, 2016. https://www.washingtonpost.com/news/worldviews/wp/2016/11/16/is-trump-a-berlusconi-let-a-berlusconi-expert-explain/?noredirect=on&utm_term=.964b761f785c.

Taylor, Victoria. 2000. *Para/Inquiry: Postmodern Religion and Culture*. London: Routledge.

Telese, Luca. 2011. "Berlusconi e la diplomazia degli insulti." *Il Fatto Quotidiano*, September 13, 2011. http://www.ilfattoquotidiano.it/2011/09/13/berlusconi-e-la-diplomazia-degli-insulti/157071/.

"Terremoto a L'Aquila, rinviata a giudizio la commissione Grandi Rischi." 2011. *Corriere della Sera*, May 25, 2011. http://www.corriere.it/cronache/11_maggio_25/terremoto-aquila-grandi-rischi-commisione-rinvio-giudizio_93393056-86c6-11e0-a06d-0594606c12ff.shtml.

Thompson, Derek. 2014. "The Algorithm Economy: Inside the Formulas of Facebook and Amazon." *Atlantic,* March 12, 2014. https://www.theatlantic.com/business/archive/2014/03/the-algorithm-economy-inside-the-formulas-of-facebook-and-amazon/284358/.

Tipaldo, Giuseppe. 2015. "Quando la scienza trema: Scienza, pseudoscienza, politica e media nel terremoto dell'Aquila." In *Terremoti, comunicazione, diritto: Riflessioni sul processo alla 'Commissione Grandi Rischi,'* edited by Alessandro Amato, Andrea Cerase, and Fabrizio Galadini, 203–220. Milano: Franco Angeli.

——. 2019. *La società della pseudoscienza: Orientarsi tra buone e cattive spiegazioni*. Bologna: Il Mulino.

Tommaseo, Nicolò, and Bernardo Bellini. 1869. *Dizionario della lingua italiana, con alter centomila giunte ai precedent dizionari*. Torino e Napoli: L'Unione tipografico-editrice Torinese.

Tondo, Lorenzo, and Katherine Leyton. 2008. "Berlusconi's Trophy Cabinet." *Globe and Mail* (Toronto), April 26, 2008.

Toselli, Paolo. 2004. *Storie di ordinaria falsità: Leggende metropolitane, notizie inventate, menzogne; I falsi macroscopici raccontati da giornali, televisioni e Internet*. Milan: Biblioteca Universale Rizzoli.

Totaro, Paolo, and Domenico Ninno. 2014. "The Concept of Algorithm as an Interpretive Key of Modern Rationality." *Theory, Culture & Society* 31 (4): 29–49. https://doi.org/10.1177/0263276413510051.

Tribunale di L'Aquila. 2013. *Sezione penale*, del 22/10/2012, n. 380 (depositata il 19/01/2013). http://www.magistraturademocratica.it/mdem/qg/doc/Tribunale_di_LAquila_sentenza_condanna_Grandi_Rischi_terremoto.pdf.

Trocchi, Cecilia Gatto. 1990. *Magia ed esoterismo in Italia*. Milan: Mondadori.

——. 1998. *Nomadi spirituali: Mappe dei culti del nuovo millennio*. Milan: Arnaoldo Mondadori Editore.

Turner, Victor, and Edith Turner. 1978. *Image and Pilgrimage in Christian Culture: Anthropological Perspectives*. New York: Columbia University Press.

van Dijk, Jan, and Kenneth L. Hacker. 2018. *Internet and Democracy in the Network Society*. London: Routledge.

van Dijck, José. 2014. "Datafication, Dataism and Dataveillance: Big Data between Scientific Paradigm and Ideology." *Surveillance and Society* 12 (2): 197–208.

Vannini, Walter. 2015. "#Algofobia: Chi ha paura dell'algoritmo cattivo?" *Tech Economy*, August 10, 2015. http://www.techeconomy.it/2015/10/08/algofobia-paura-dellalgoritmo-cattivo/.

"Venerdi 17: La Giornata anti-superstizione del CICAP." CICAP.org. https://www.cicap.org/n/articolo.php?id=275351

Walsh, Lynda. 2013. *Scientists as Prophets: A Rhetorical Genealogy*. New York: Oxford University Press.

"Walter Ricciardi: 'Le dimissioni dall'Iss? Da governo posizioni antiscientifiche." 2019. *Il Fatto Quotidiano*. https://www.ilfattoquotidiano.it/2019/01/02/walter-ricciardi-le-dimissioni-dalliss-da-governo-posizioni-antiscientifiche/4869877/

Watters, Clare. 2011. "Being Berlusconi: Sabina Guzzanti's Impersonation of the Italian Prime Minister between Stage and Screen." In *Between Political Critique and Public Entertainment*, edited by Villy Tsakona and Diana Elena Popa, 167–189. Amsterdam: John Benjamins.

Weber, Peter. 2016. "Stephen Colbert resurrects his *Colbert Report* 'The Word' segment to define 'Trumpiness.' *The Week*. *https*://theweek.com/speedreads/636881/stephen-colbert-resurrects-colbert-report-word-segment-define-trumpiness

Wedeen, Lisa. 1999. *Ambiguities of Domination: Politics, Rhetoric, and Symbols in Contemporary Syria*. Chicago: University of Chicago Press.

Weseltier, Leon. 2015. "Among the Disrupted." *New York Times*, January18, 2015. http://www.nytimes.com/2015/01/18/books/review/among-the-disrupted.html?ref=todayspaper.

West, Sarah Myers. 2017. "Data Capitalism: Redefining the Logics of Surveillance and Privacy." *Business & Society* 58 (1): 20–41. https://doi.org/10.1177/0007650317718185.

Weston, Kath. 2017. *Animate Planet: Making Visceral Sense of Living in a High-Tech Ecologically Damaged World*. Durham, NC: Duke University Press.

Wilf, Eitan. 2013. "Toward an Anthropology of Computer-Mediated, Algorithmic Forms of Sociality." *Current Anthropology* 54 (6): 716–739. https://doi.org/10.1086/673321.

Wilken, Rowan. 2011. "Fantasies of Control: *Numb3rs,* Scientific Rationalism, and the Management of Everyday Security Risks." *Continuum: Journal of Media & Cultural Studies* 25 (2): 201–211. https://doi.org/10.1080/10304312.2011.553941.

Winner, Langdon. 1986. *The Whale and the Reactor: A Search for Limits in an Age of High Technology*. Chicago: University of Chicago Press.

World Skeptics. 2012. "6th World Skeptics Congress: Promoting Science in an Age of Uncertainty." *World Skeptics Organization*. http://www.worldskeptics.org./about-us.

Yang, Wesley. 2017. "Is the 'Anthropocene' Epoch a Condemnation of Human Interference—or a Call for More?" *New York Times Magazine*, February 14, 2017. https://www.nytimes.com/2017/02/14/magazine/is-the-anthropocene-era-a-condemnation-of-human-interference-or-a-call-for-more.html.

Yeo, Michael. 2014. "Fault Lines at the Interface of Science and Policy: Interpretive Responses to the Trial of Scientists in L'Aquila." *Earth-Science Reviews* 139:406–419. https://doi.org/10.1016/j.earscirev.2014.10.001.

Yurchak, Alexei. 1997. "The Cynical Reason of Late Socialism: Power, Pretense, and the *Anekdot*." *Public Culture* 9:161–188. https://doi.org/10.1215/08992363-9-2-161.

Zammito, John. 2004. *A Nice Derangement of Epistemes: Post-positivism in the Study of Science from Quine to Latour*. Chicago: University of Chicago Press.

Žižek, Slavoj. 1989. *The Sublime Object of Ideology*. London: Verso.

——. 2006. *The Parallax View*. Cambridge, MA: MIT Press.

——. 2009. "Berlusconi in Tehran." *London Review of Books* 31 (14): 3–7. http://www.lrb.co.uk/v31/n14/slavoj-zizek/berlusconi-in-tehran.

Zocchi, Alessandro. 2011. "Psicologia della superstizione: Perche si mantengono i comportamenti inutile?" *Cicap.org*. http://www.cicap.org/new/stampa.php?id=101608.

Zuboff, Shoshana. 2015. "Big Other: Surveillance Capitalism and the Prospects of an Information Civilization." *Journal of Information Technology* 30:75–89.

Index

Note: Page references in *italics* refer to illustrative matter.

absurdity, 159–60
Accademia Nazionale dei Lincei, 97–99
Agamben, Giorgio, 71, 110
Alberti, Massimo, 142–43, 164
Alcaro, Riccardo, 74
Algorithmic Culture (Striphas), 156–57
algorithms: art and beauty of, 173n7; capitalism of, 146–47, 158–59, 179n13; culture of, 76, 178n7, 178n9; democracy through, 93–95; development of, 177n1 (concl.); digital personalization through, 146–54, 178nn3–5; for disaster management, 174n17; filter bubbles, 154–56; manufacturing metaphor and, 178n6; mystification of, 156–59, 179n10; rise of populism and, 73–77, 163–64. *See also* data surveillance; digital populism; Rousseau (digital platform)
The Algorithms of Capital (Harney), 73, 153–54
Alleanza Nazionale (AN), 34
Allegranti, David, 87
alternative medicine, 57–59
Amato, Giuliano, 21
Amazon, 76, 152
Amodeo, Francesco, 82
anesthetization, scientific, 13, 105–6, 139, 144–45. *See also* risk communication; trust
animal trope, 85
Anonymous, 152
Anthropocene, 24, 122–28, 132–36, 177n2
"Anthropocene, the Year Man Changed the Planet" (Berberi), 126
antiscience conspiracies, 87–88. *See also* conspiracy theories
Aristotle, 96
The Atlantic (publication), 152

Bailey, Michael, 60
Barberi, Franco, 105, 107
Bartie, Whitney, 102–3
Baudo, Pippo, 79
behind-ism, 130–31, 132, 136–40, 177n5
belief, 10, 108–9, 112–13, 158–59
The Beppe Grillo Show (television show), 79
Berberi, Leonard, 126
Berlusconi, Silvio, 31, *32*; context of political rise of, 34, 41–43, 169n4, 170n10; humor and, 32, 35; media companies of, 22, 35, 170nn14–15; mediatized politics of, 21–23, 27, 167n10; objectification of women by, 20, 32, 35, 39, 168n11; reputation of, 3–4, 18, 36, 168n13, 169n3; response to earthquake, 116–17; scandals and imprisonment of, 20, 21, 31–32; *Una storia italiana,* 35, 41, 162; terms of, 2, 34–35, 165n1; underage prostitution and sex parties by, 32, 35, 45, 77, 86, 114, 169n13, 170n17; as zombie, 77. *See also* Forza Italia (FI)
Bersani, PierLuigi, 20
Bersani, Walter, 77
Bertolaso, Guido, 117–18
Besteman, Catherine, 84–85, 93, 179n11
Billi, Marco, 105, 107, 175n2
Billig, Michael, 37
Billionaires (protest group), 50
Bin Laden Can Get on TV, but I Can't (DVD), 43
Biorcio, Roberto, 78
BioSciences (publication), 56
Bloomberg, David, 55, 56
Bogost, Ian, 156, 157, 178n6
Boldrini, Laura, 16
Bordignon, Fabio, 78
Boschi, Enzo, 118
Boyer, Dominic, 13, 20
Bozdag, Engin, 155–56
Brewer, Richard, 56
Brozo (character), 171n18
"bunga bunga," 45, 170n17
Bush, George W., 160, 165n3, 167n7, 169n8
Butler, Ella, 71

Calcutt, Andrew, 7
Cambridge Analytica, 151
capitalism, 14–15, 54, 79, 158–59, 179n13
Casaleggio, Gianroberto, 88, 173n12, 174n13, 177n1 (concl.)
Casaleggio Association, 73
Cassandra's prediction, 26, 129–30
Castells, Manuel, 11, 15
"The Cathedral of Computation" (Bogost), 156
Catholic Church, 3, 14, 29, 34, 53–56, 61–62, 69, 171n4, 171n6
causality of disaster, 121–22, 126–28. *See also* Anthropocene; climate change; disaster prediction; National Commission for the Forecast and Prevention of Major Risks (CGR)
Ceccarini, Luigi, 78
censorship, 43
CGR. *See* National Commission for the Forecast and Prevention of Major Risks (CGR)
chemtrails, 30, 133–34, 136. *See also* geological conspiracy theories
Cheney-Lippold, John, 147
Chile, 103
China, 135
Christian Democrats (DC), 34, 40
Cialente, Massimo, 118, 175n6
CICAP (Committee for the Investigation of Pseudoscientific Claims), 27–28, 144; Day against Superstition by, 1–2, 27, *53,* 55, 59, *64,* 65, 71–72; Day for the Correct Scientific Information, 57; as a part of the skepticism and rationalist movements, 4, 9, 13, 17–18, 20, 55–57, 63, 71
Ciccozzi, Antonello, 108–10, 120
cinepanettone, 38–39
Cingolani, Sofia, 81–83, 173n9
citizen-subject, truth and governance of, 7–8
Civil Protection Department (DPC), 105, 116, 117–18, 129, 137–38, 140, 177n21
cleverness, 18, 36, 38, 40, 45, 63, 78, 82, 86
climate change: causation of, 12, 104, 125, 132, 133; denial of, 128, 177n3; Pope Francis on, 126–27. *See also* Anthropocene
cloud seeding, 135
clown metaphor, 36, 38, 163, 169n14, 171n18. *See also* comedy-style of Italy, as genre
Colbert, Stephen, 10, 34, 165n3
The Colbert Report, 39, 50

Comaroff, Jean and John, 14, 54, 166n7, 176n15, 179n14
comedy-style of Italy, as genre, 38–39, 169n7. *See also* news parody
Comitato Italiano per il Controllo delle Affermazioni sul Paranormale. *See* CICAP
Comitato italiano per l'investigazione del pseudoscienza. *See* CICAP
Commissione Grandi Rischi. *See* National Commission for the Forecast and Prevention of Major Risks (CGR)
Committee for Skeptical Inquiry (CSI), 171n3
Committee for Skeptical Inquiry in the United States, 55
Committee for the Investigation for Pseudoscientific Claims. *See* CICAP
Committee for the Investigation of Claims of the Paranormal. *See* CICAP
Committee for the Scientific Investigation of Claims of the Paranormal (CSICOP), 171n3
Communist Party (PCI), 34, 40, 169nn4–5
conspiracy theories, 29–30, 87–88, 131–36, 174n13
Conte, Giuseppe, 21
Coppi, Franco, 118
Corriere della Sera (publication), 81
Couldry, Nick, 146, 153
Cracking Up (Lewis), 41
Craxi, Bettino, 40–42, 43; Grillo on, 78, 79
Creation Museum, 71
Critchley, Simon, 158
Crutzen, Paul, 123
Curtis, Neal, 15
cynicism, 37–40, 174n13. *See also* distrust; humor; skepticism; trust

The Daily Show, 39, 50
D'Alema, Massimo, 21
D'Alimonte, Roberto, 75
Dataclysm (Rudder), 149–50
datafication, 12, 155, 166n6, 178n8. *See also* algorithms; scientific truth
data surveillance, 30, 146–49. *See also* algorithms; filter bubbles
Day against Superstition, 1–2, 27, *53,* 55, 59, *64,* 65, 71–72. *See also* CICAP (Committee for the Investigation of Claims of the Paranormal)
Day for the Correct Scientific Information, 57. *See also* CICAP (Committee for the Investigation of Pseudoscientific Claims)
DC (Democrazia Cristiana), 34, 40
De Bernardinis, Bernardo, 109
Deleuze, Gilles, 37
De Martino, Ernesto, 60, 68, 112–13, 176n14
Democratic Party (PD), 21, 34, 73–74, 77, 82, 90, 172n3
Democratic Party of the Left (PDS), 34, 169n5
Democrazia Cristiana (DC), 34, 40
denialism, 87
Department of Civil Protection. *See* Civil Protection Department (DPC)
dice game, 65–66, 68
digital populism, 4, 5, 26, 30, 79–83, 93–95. *See also* algorithms; Five Star Movement (M5S)
digital surveillance, 30, 146–49. *See also* algorithms
Di Grazia, Salvo, 57–58
Di Maio, Luigi, 26, 74, 92–93, *155,* 161, 163–64, 172n4, 174n16
Dini, Lamberto, 21
DiPaolo, Marc, 173n10
Dipartimento di Protezione Civile. *See* Civil Protection Department (DPC)
disaster management, 114–17, 174n17
disaster prediction, 2–3, 26, 101–4, 107, 111–12, 118, 128–31, 174n17. *See also* earthquakes; National Commission for the Forecast and Prevention of Major Risks (CGR); risk communication
disease, conspiracy theories on, 87, 173n11
disinformation, 15–18, 134, 166n5. *See also* truth, as social production
disinformation society, as term, 15, 17. *See also* post-truth era
Disinformazia (Nicodemo), 133
"Dissertation on the Origin and Foundation of the Inequality of Mankind" (Rousseau), 92
distrust, 20, 37, 49, 81, 91–92, 167n9. *See also* cynicism; skepticism; trust
divination, 63, 101, 172n7
docu-media, 14
Draquila (film by Guzzanti), 114, 117, 176n17
Drive In (television show), 42
Dusi, Elena, 126

earthquakes: (2009) L'Aquila, 2, 25, 96, 175n1; causality of, 121, 132, 135; destruction of, 96–97, 120, 141; rebuilding after, 97, 116–17, 136–37, 176n11; risk communication on, 2, 104–5, 122, 128–30, 134–35, 174n17, 177n4; trial in Turkey for, 102. *See also* disaster management; disaster prediction; National Commission for the Forecast and Prevention of Major Risks (CGR)
Ecce Bombo (film), 169n7
economic uncertainty, 20, 53
Economist (publication), 20
empirical reality production, 6
enchantment trope, 85–87. *See also* zombie imaginaries
entrepreneurial populism, 37
epistopolitics, 23–26
The Era of Post-truth (Cosentino), 133
erosion of facts, history of, 6–10. *See also* disinformation; truth, as social production
European Council of Skeptical Organizations, 55
European Court of Human Rights, 101–2
The European Matrix (Amodeo), 82
Eva, Claudio, 106
exceptional law, 116–17, 176n10, 177n20

Facebook, 89, 91, 149–52, 155–57
factish, as term, 9, 10
facts: history of erosion of, 6–10; obscuring of, 15
The Faith of the Faithless (Critchley), 158
fake medicine, 57–59
"fake news": disinformation as, 16–20; Pope Francis on, 1, 16
Fantastico (television show), 43, 79
Fede, Emilio, 45
Federal Tort Claims Act (FTCA), 102
Fernandez, James W., 34
Ferraris, Maurizio, 13–14, 22
Fieschi, Catherine, 37
film genre, 38–39
The Filter Bubble (Pariser), 154
filter bubbles, 154–56. *See also* algorithms; data surveillance
Fine, Gary, 8, 102
Fininvest, 34, 42
Five Star Movement (M5S): on EU membership, 174n15; Grillo and political positioning of, 20, 21, 24–26, 28, 88–93, 172n3; Renzi's criticism on, 73; rise of algorithm populism of, 4, 5, 73–77, 81, 93–95, 144. *See also* Grillo, Beppe; Rousseau (digital platform)
flooding, 102
Fo, Dario, 43
F—Off Day, 80
Forbes (publication), 79
Forza Italia (FI), 21, 34, 35, 73, 74, 162. *See also* Berlusconi, Silvio
Foucault, Michel, 7–8, 159
Francis (pope), 1, 16, 126–28
Francis of Assisi (saint), 92, 174n12
Friday the 17th, as unlucky, 2, 3, 65
Fulginiti, Valentina, 22
furbo, 18

Gabibbo (character), 27, 36, 47–49, 171n18. See also *Striscia la Notizia* (television show)
Gargarello, Romano, 2
Gehl, Robert, 122, 130
geological conspiracy theories, 29–30, 131–33. *See also* conspiracy theories; scientific truth
giant warming machines, 30. *See also* geological conspiracy theories
Gilmour, David, 31, 49
Ginsborg, Paul, 35, 168n13
Giugno, Linda, 175n4
Giuliani, Giampaolo, 128–31, 177n4
Giuliani, Giuliano, 26
Glaeser, Andreas, 4
Go Italy (FI), 21, 34
Golden Tapir Award, *44,* 45
Google, 89, 148, 150, 154–55, 157, 174n13, 178n3, 178n6
"Google Knows What You're Doing Tomorrow" (Oremus), 150
Gordon, Avery, 85
Graeber, David, 166n7
Gramsci, Antonio, 42
Gran Sasso National Laboratory, 120–21, 128–31
Grant, Rachel, 103
Great Risk Commission. *See* National Commission for the Forecast and Prevention of Major Risks (CGR)
Greggio, Ezio, 42–43

Grillo, Beppe, *76*; antiscience conspiracies by, 87; on Craxi, 43; criminal charges against, 173n8; election of, 20, 74, 77, 91, 172n3; *Modern Slaves,* 79–80; political strategy of, 18, 28; zombie imaginaries of, 77–78, 83–87. *See also* Five Star Movement (M5S); Rousseau (digital platform)
Grosz, Elizabeth, 73
Guardian (publication), 103
Guinness Book of World Records, 46
Gusterson, Hugh, 84–85, 147–48, 157, 158, 179n12
Guyer, Jane, 177n2
Guzzanti, Sabina, 43, 114, 117

HAARP (High Frequency Active Auroral Research Program), 133–34, 177n6
Halpern, Michael, 97
Haraway, Donna, 69
Hardt, Michael, 54
Harney, Stefano, 73, 153–54
Hasian, Marouf, Jr., 122, 130
Haugerud, Angelique, 50
Head, Lesley, 123
Helmreich, Stefan, 124
Herzfeld, Michael, 35
Hetherington, Kregg, 11, 166nn4–5
Heywood, Paul, 37
High Frequency Active Auroral Research Program (HAARP), 133–34, 177n6
homeopathic medicine, 57–58
Huber, Mary Taylor, 34
humanism, 61
humor, 32, 35, 37–40, 169n8. *See also* comedy-style of Italy, as genre
Hurricane Audrey (1957), 102–3
hurricanes, 102–3

Iacchetti, Enzo, 45
Le Iene (television show), 161, 179n15
I Exist (film), 142–43
I'll Give You America (television show), 79
illusionism, 62–63
Imaging Cosmic and Rare Underground Signals (ICARUS), 121
informationalism, 11, 145–46
INGV (Italy's National Institute of Geophysics and Volcanology), 129
Instituto Superiore di Sanità (ISS), 87
International Commission on Earthquake Forecasting for Civil Protection (ICEFCP), 107
International Union of Geodesy and Geophysics, 97
Internet users, statistics on, 11
Io C'é (film), 142–43
Ionospheric Research Instrument (IRI), 133–34
Iovino, Serenella, 99
irony, 34. *See also* humor
Istituto Nazionale di Geofisica e Vulcanologia (INGV), 129
Italian comedy, as genre, 38–39, 169n7
Italian Communist Party (PCI), 34, 40, 169nn4–5
Italian Democratic Party, 21, 34, 73–74, 77, 90, 172n3
Italian Enlightenment, 61, 91–92
Italian Socialist Party (PSI), 40, 42–43, 79
Italozombies, 83–87. *See also* zombie imaginaries
Italy's National Institute of Geophysics and Volcanology (INGV), 129

Jameson, Frederic, 7, 135
Jarry, Alfred, 70, 159–60
jettatore, 1, *5,* 65, 112–13, 176n12
Jones, Erik, 35
Jones, Graham, 62
The Joy of the Gospel (Pope Francis), 128
junk science, 57–58

Kamman, Richard, 68
Kenesson, Frank, 160
Kennedy, Paul, 166n7
knowledge: *vs.* belief, 10, 158–59; mediatization of, 11–16. *See also* scientific truth
Kohn, Eduardo, 123
Koopman, Colin, 142, 155

Laboratory Life (Latour), 8
Labor Disorders in Neoliberal Italy (Molé), 108
L'Anomalo bicefalo (television show), 43
L'Aquila, Italy, 2–3, 25, 96, 104, *115,* 118–19. *See also* National Commission for the Forecast and Prevention of Major Risks (CGR)
The L'Aquila Earthquake (publication), 2

La Repubblica (publication), 81, 105
Latour, Bruno, 1, 7, 8–10, 15, 104, 123–24, 158
Laudato Si' of the Holy Father Francis on Care for Our Common Home (Pope Francis), 126
Lega Nord (LN), 21, 34, 74, 172n5
Leone, Giovanni, 169n3
Le Pen, Jean-Marie, 169n2
Lesher, Alan, 97, 175n23
L'Espresso (publication), 132
Lessidri, Giacomo, 63
Letta, Enrico, 20, 74, 172n3
Lewis, Paul, 41
The License (play by Pirandello), 112, 176n16
Life by Algorithms (Besteman and Gusterson), 84–85, 147
LinkedIn, 148
lottizzazione, 42, 80
Lukes, H. N., 160
Luttazzi, Daniele, 43
Lyotard, Jean-Francois, 7

M5S. *See* Five Star Movement (M5S)
MacKenzie, Adrian, 150–51
magic, 14, 60, 62–63. *See also* superstition
Major Risk Commission. *See* National Commission for the Forecast and Prevention of Major Risks (CGR)
Mamma Mia: Berlusconi Explained for Posterity and Friends Abroad (Severgnini), 49–50
manufacturing metaphor, 178n6
Mapei, Camillo, 59
Marcus, George E., 17, 167n8
Marks, David, 68
Marshall, Jonathan, 15
Marxism, 14, 34, 42, 153
Massumi, Brian, 175n9, 176n19
Match.com, 150
Mbembe, Achille, 37, 38, 45, 50
McGuire, Bill, 125
McNally, David, 179nn13–14
McQuillan, Don, 151
Meana, Marina Ripa di, 40–41
Mediaset, 22, 35, 43, 162. *See also* Berlusconi, Silvio
mediatized politics, 11–16, 22–23, 42–43, 80–81, 144
medicine, 57–58
Melanco, Mirco, 169n7
Merkel, Angela, 26, 32
Merlo, Francesco, 38
Minuz, Andrea, 93, 161
mirrors: algorithms and online user activity as, 148–49; Grillo on, 85, 142; selfism's use of, 142–43; superstition of shattering, 27, *64,* 67, 68
Mishra, Pankaj, 92
Misinformation (Quattrociocchi and Vicini), 133
Mittelstrass, Jürgen, 13
Modern Slaves (Grillo), 79–80
Moe, Nelson, 61–62
Mohammed, Shaheed, 166n5
Monti, Mario, 74, 77
Movimento Cinque Stelle. *See* Five Star Movement (M5S)
Mussolini, Benito, 31, 32

Natale, Paolo, 78
National Alliance (AN), 34
National Commission for the Forecast and Prevention of Major Risks (CGR): aftermath of trial, 118–19; charges and trial against, 2–3, 17–18, 24, 28–29, 97, 105–12, 175n2; cultural context of the trial against, 97–100, 112–18, 145, 176nn15–16; risk communication by, 2, 104–5, 122, 129–30, 175n8. *See also* disaster prediction; earthquakes
National Healthcare Service, 110
National Health Institute (ISS), 87
National Institute of Geophysics and Volcanology (INGV), 118, 129
National Weather Bureau (U.S.), 102–3
natural disasters, 100. *See also* disaster management; disaster prediction; earthquakes
neoliberalism, 15, 54, 79, 169n14, 179n12
New England Skeptical Society, 55
Newitz, Annalee, 85
The News is Creeping news program, 27, 32, 79
news parody, 39–40, 42–43, 170n9. *See also* comedy-style of Italy, as genre; *Striscia la Notizia* (television show)
Newsweek, 7
New York Times (publication), 20, 150, 166n3
Ninno, Domenico, 153

Northern League (LN), 21, 34
Novella, Steven, 55, 56
Numb3rs (television show), 71

Obama, Barack, 32, 166n3, 173n10
objectification of women, 20, 32, 35, 39, 168n11, 170n16. *See also* prostitution and sex parties
occultism, 62, 158–59, 179n13. *See also* magic; superstition; zombie imaginaries
oddmatch, 68
Odiens (television show), 43, 170n13
On the Modern Cult of the Factish Gods (Latour), 9
oracles, 101
Oremus, Will, 150
Organization for the Eradication of Superstition (ANIS), 55
Orsi, Robert, 69
Orsina, Giovanni, 167n10
Oscillations Project with Emulsion-tRacking Apparatus (OPERA), 121, 177n1 (ch. 5)

Paliewicz, Nicholas, 122, 130
para, as prefix, 16–17
Para/Inquiry (Taylor), 17
parasite, 18, 167n8
para-site, as term, 18
Para-Sites (Marcus), 17
Pariser, Eli, 154, 178n3
Parmalat, 79
Partito Comunista Italiano (PCI), 34, 40, 169nn4–5
Partito Democratico (PD), 21, 34, 73–74, 77, 90, 172n3
Partito Democratico della Sinistra (PDS), 34, 169n5
Partito Radicale (PR), 81
Partito Socialista Italiana (PSI), 40, 42–43, 79
La patente (play by Pirandello), 112
patrimonial authority, 35–36
PCI. *See* Communist Party (PCI)
PD. *See* Democratic Party (PD)
PDL. *See* The People of Freedom Party (PDL)
PDS. *See* Partito Democratico della Sinistra (PDS)
The People of Freedom Party (PDL), 31, 170n10, 172n3
personalization, 146–54. *See also* algorithms
The Philosophy of Berlusconi (Solla), 70
pig trope, 85
Pirandello, Luigi, 112, 176n16
placebo effect, 58
Polidoro, Massimo, 2, 53, 59, 171n1
Polverini, Renata, 44
Poovey, Mary, 9
Il Popolo della Liberta (PDL), 31, 170n10, 172n3
populism: algorithms and digital populism, 4, 5, 26, 30, 73–77, 79–83, 93–95, 163–64; entrepreneurial populism, 37; showman populism, 46–47
postmodernism theory, 7
poststructuralism, 7
post-truth era, 6–10, 166n4. *See also* disinformation society, as term; "fake news"
probability, 65–69, 144
Prodi, Romano, 21, 35
"the projector" *(jettatore),* 1, 5, 65
prophecy *vs.* science, 103–4
prostitution and sex parties, 32, 35, 45, 77, 86, 114, 169n13. *See also* objectification of women
Protestantism, 60, 61, 69
Publitalian, 22. *See also* Berlusconi, Silvio
puppets. *See* Gabibbo (character)
purity myth, 92

Quack, Johannes, 55

Radical Party (PR), 81
radiestesia, 63
Radiotelevisione Italiana (RAI), 42, 79
radon emissions and earthquake predictions, 26, 103, 128–31
Raffnsøe, Sverre, 124–25
rainstorms, 135
RaiOt (television show), 43
Rasmussen, Anders Fogh, 32
Rational Examination Association of Lincoln Land, 55
rationalism, 55, 71. *See also* skepticism
RC (Rifondazione Comunista), 169n5. *See also* PCI
reactionary rationalism, 118–19
religion, 60, 171n6. *See also* Catholic Church; selfism

Renan, Ernest, 60, 69
Renzi, Matteo, 73, 172n2, 173n6
Repubblica Sociale party, 32
Republic Giorgio Napolitano, 97
Rete 4, 35
rettore, 2, 165n2
Ricci, Antonio, 42–43, 48, 79. See also *Striscia la Notizia* (television show)
Ricciardi, Walter, 87, 173n11
Rifondazione Comunista. *See* RC
right-wing antiestablishment populism. *See* Five Star Movement (M5S)
Risk Commission. *See* National Commission for the Forecast and Prevention of Major Risks (CGR)
risk communication, 2, 104–5, 122, 132–33, 137–40. *See also* disaster prediction; National Commission for the Forecast and Prevention of Major Risks (CGR); scientific anesthetization; scientific prediction; trust
ritualized magic and superstition, 16, 61–62, 66–69, 108, 166n7
Robert-Houdin, Jean-Eugène, 63
Roitman, Janet, 38, 50
Rousseau (digital platform), 73–74, 88–95, 174n14. *See also* Casaleggio, Gianroberto; Five Star Movement (M5S)
Rousseau, Jean-Jacques, 91–92, 121
Rudder, Christian, 149–50
Russia, 102, 135

salt, 65–66, 68
Salvini, Matteo, 74, 87, 172n4
satire. *See* news parody
Scalfari, Eugenio, 31, 80
Schroeder, Gerhard, 32
Schüll, Natasha, 147
sciechimiche, 30, 133–35, 136
scientific activism, 53–54, 166n3
scientific anesthetization, 13, 105–6, 139, 144–45. *See also* risk communication; trust
The Scientific Life (Shapin), 20
scientific prediction, 2–3, 101–4, 107, 111–12, 118. *See also* National Commission for the Forecast and Prevention of Major Risks (CGR)
scientific skepticism, 55–57, 171n3. *See also* skepticism
scientific truth, 3–6; behind-ism and, 130–31, 132, 136–40; competition with paratruths, 17; Fine on, 8; Grillo's antiscience conspiracies against, 87; Lyotard on, 7; *vs.* prophecy, 103–4; skepticism and, 12–13, 53–57. *See also* knowledge; National Commission for the Forecast and Prevention of Major Risks (CGR); truth as social production
Scientists as Prophets (Walsh), 103
Seaver, Nick, 157, 178n9
selfies, 148–49, 156
selfism, 142–43
Selvaggi, Giulio, 97
Semblance and Event (Massumi), 175n9, 176n19
Serres, Michael, 167n8
Severgnini, Beppe, 49–50
sex parties and underage prostitution, 32, 35, 45, 77, 86, 114, 169n13, 170n17. *See also* objectification of women
Shapin, Steven, 8–9, 19–20, 167n9
sharing-out media system, 42, 80
Shklar, Judith, 122
showman populism, 46–47
Skeptical Inquirer (publication), 171n3
skepticism, 12–13, 53–59, 171n3. *See also* cynicism; trust
Sky Italia, 43
A Social History of Truth (Shapin), 8–9
Socialist Party (PSI), 40, 42–43, 79
Social Republic party, 32
social theory of knowledge, 6
Solla, Gianluca, 70–71
Sordi, Alberto, 39
Stewart, Jon, 50
Stoermer, Eugen, 123
Una storia italiana (Berlusconi), 35, 41, 162
Stories of Ordinary Falsehood (Toselli), 133
Striphas, Ted, 76, 156–57, 178n7
Striscia la Notizia (television show), 32; audience reviews of, 46–47, 49; awards of, 46, 48; cultural context and rise of, 27, 34, 40; description of, 43–46; Gabibbo as mascot of, 27, 36, 47–49, 171n18; production of, 79, 170n12; translations of title of, 169n6
Sulmona, Italy, 129, 131
super, as prefix, 17

superstition, 54, 59–69, 171n5. *See also* magic
surveillance, 30, 146–49. *See also* algorithms
swarms (small tremors), 2, 25, 104, 107, 119, 130, 175n3. *See also* earthquakes

Tangentopoli trials, 41, 42, 79
Taylor, Victor E., 17
television to Internet shift in media consumption, 24–25
theatrical political activism, 32–34
The Web Is Dead, Long Live the Web (Casaleggio), 88
Thompson, Derek, 152–53
Time (publication), 79, 135, 146
toads, 103
Totaro, Paolo, 153
tremors. *See* swarms (small tremors)
Trocchi, Cecilia Gatto, 62
Trump, Donald, 10, 14, 151, 154, 166n3, 168n13
Trumpiness, as term, 10
trust: in algorithms, 85, 90–93, 155; morality and, 9; in science, 13, 18–20, 105–6, 113, 130–32, 138–39, 144–45, 166n6; in *Striscia* and Gabibbo, 46–49. *See also* cynicism; risk communication; skepticism
truth as social production, 7–10, 165n3, 166nn4–5. *See also* scientific truth
truthiness, as term, 10, 165n3
truth-tellers, 19–20. *See also* Gabibbo (character)
Turkey, 102
Turner, Edith and Victor, 69

ubu-esque power, 70, 159–60, 163
Ubu Roi (play by Jarry), 70, 159–60
Union of Concerned Scientists, 97
United States: climate change denial in, 177n3; politics and truth in, 165n3, 166n4, 166n7; presidential elections, 41, 166n3, 169n8; superheroes and zombie narratives in, 173n10; weather and disaster management in, 102–3, 132, 135

vaccination, 87, 173n11
Vaffa Day, 80
vampire trope, 114–16
Vanderbilt, Cornelius, 14
V-Day, 80
veline, 44, 170n16
Videobox (television show), 79
Vittorini, Vicenzo, 105
Voltaire, 60, 91, 121

Waking the Giant (McGuire), 125
Walsh, Lynda, 103–4, 127–28
water divination, 63, 172n7
wealth, 14, 54. *See also* capitalism
Webegg, 88, 177n1 (concl.)
Weber, Max, 35–36, 104
Wedeen, Lisa, 38
We Have Never Been Modern (Latour), 158
Weseltier, Leon, 178n8
Weston, Kath, 125
WikiLeaks, 152
Wilken, Rowan, 71
wireless phone users, statistics on, 11
witch trials, 61
"Witchnight" (La Notte della Streghe), 52–53
The Wizard of Oz, 10
World Communications Day, 1, 16
World Skeptics Congress, 55

yellow jackets, 81
Yurchak, Alexei, 38

Zacchero, Vincenzo, 44
Zapponi, Carlo, 173n7
Žižek, Slavoj, 31, 37, 105–6, 175n7
zombie imaginaries, 77–78, 83–87, 158–59, 173n10, 179n13
Zuckerberg, Mark, 151. *See also* Facebook

CPSIA information can be obtained
at www.ICGtesting.com
Printed in the USA
LVHW032023280622
722267LV00003B/261

9 781501 750793